PERFORMING DECEPTION

Performing Deception

Learning, Skill and the Art of Conjuring

Brian Rappert

https://www.openbookpublishers.com

© 2022 Brian Rappert

This work is licensed under a Creative Commons Attribution-NonCommercial 4.0 International (CC BY-NC 4.0). This license allows you to share, copy, distribute and transmit the text; to adapt the text for non-commercial purposes of the text providing attribution is made to the authors (but not in any way that suggests that they endorse you or your use of the work). Attribution should include the following information:

Brian Rappert, *Performing Deception: Learning, Skill and the Art of Conjuring*. Cambridge, UK: Open Book Publishers, 2022, https://doi.org/10.11647/OBP.0295

In order to access detailed and updated information on the license, please visit, https://doi.org/10.11647/OBP.0295#copyright

Further details about the CC BY-NC license are available at https://creativecommons.org/licenses/by-nc/4.0/

All external links were active at the time of publication unless otherwise stated and have been archived via the Internet Archive Wayback Machine at https://archive.org/web

Updated digital material and resources associated with this volume are available at https://doi.org/10.11647/OBP.0295#resources

ISBN Paperback: 9781800646902
ISBN Hardback: 9781800646919
ISBN Digital (PDF): 9781800646926
ISBN Digital ebook (EPUB): 9781800646933
ISBN Digital ebook (AZW3): 9781800646940
ISBN XML: 9781800646957
Digital ebook (HTML): 9781800646964
DOI: 10.11647/OBP.0295

Cover image: Follower of Michelangelo Merisi da Caravaggio, *The Card Players* (17th century). Harvard Art Museums/Fogg Museum, Friends of the Fogg Art Museum Fund, Photo ©President and Fellows of Harvard College, 1929.253, https://hvrd.art/o/231633.
Cover design by Anna Gatti

Contents

Preface: Attention, Attention, Attention! — vii
Transcription Notes — xiii
1. A Kind of Magic — 1
2. Self and Other — 21
3. Control and Cooperation — 47
4. Natural and Contrived — 73
5. Proficiency and Inability — 99
6. Truth and Deception — 125
7. Control and Care — 147
8. Learning and Unlearning — 175
Index — 191
Bibliography — 195

Preface

Attention, Attention, Attention!

Crafting a preface is a delicate matter as they often seek to achieve multiple aims. One is to entice. Prefaces can seek to convince readers that what follows is interesting or important. In doing so, they aim to direct readers' attention.

The topic for this book—entertainment magic (or 'modern conjuring')—is itself an activity of directing attention. Through hand gestures, bodily movements, verbal patter and much more besides, magicians endeavor to draw attention to some matters (a coin in the right hand) whilst directing attention away from others (what is happening in the left hand). The sense of wonder, bafflement and surprise generated from what is perceived invites us to reconsider how we come to understand the world.

In particular, I consider the forms of practical reasoning and embodied skills associated with modern Western forms of entertainment magic. Learning magic is, in itself, a form of self-directed attention. Learners must understand how to comport themselves appropriately. More than just a process of disciplining certain choreographed movements, though, learning magic entails attending to others' experiences. In this sense, the simulations and dissimulations of magic can be approached as acts of regard. In *Performing Deception*, the tensions and contradictions associated with determining how to act through acknowledged trickery serve as bases for reimagining how we interact together more generally.

Another common function of a preface is to explain the impetus for a volume. As I came to appreciate, origin accounts are commonplace in conjuring. In writing about their lives and work, magicians often identify a key moment that spurred their initial curiosity. Childhood experiences of a relative performing a trick, for instance, would be a typical origin story.

In this spirit, I will offer a backstory for *Performing Deception*. This book has its origins in a public talk I attended in 2002. As part of a *Café Scientifique* series designed to promote public interest in science, a local pub in Nottingham hosted a seasoned magician who spoke about his work to debunk psychics. He began by announcing that, right before our very eyes, he would perform remarkable 'feats of the mind'—the bending of spoons with the slightest of touches, the reading of audience members' thoughts, the adding together of numbers faster than a calculator, and so on. As these feats were done, he recounted how those with malicious aims used such acts to prey on the gullible and vulnerable. Once finished, the magician then meticulously revealed, one-by-one, how each of the effects we witnessed had been accomplished.

The conclusion of these exposés seemed plain: no special powers were necessary to undertake apparently extraordinary acts.

An intermission followed.

After the break, the speaker came back to disclose that, actually, he had not accomplished the feats as suggested. He then went on to meticulously show, one-by-one, how each had actually been done. Without driving the point home in a manner that might make the audience of academics, technical professionals and scientifically inclined members of the public uncomfortable, the conclusion seemed plain enough: anyone can be fooled, you included.

What stayed with me afterwards was not the explanation of the effects. Instead, the lasting impression was the manner in which we as the audience were moved from being spectators-turned-confidants to shared (but still secret) truths, to instead being spectators-turned-played-dupes. The effect of this performance on me was long-lasting and generative.

Shortly after the talk a doubt crept into my head: had we, after all, really been shown how the bending of spoons, the reading of minds and so on had been done in the second act? Were, perhaps, the revelations behind the revelations just another staging?

As I continued to reflect on the show, later details of the event came to mind that made it difficult to square with the idea of full disclosure. For instance, the performer announced the presence of a member of the

Magic Circle, a society for magicians which I would later be accepted into in 2021. Purportedly, this person was invited to ensure that the secrets of the profession were not unduly disclosed. But had not quite a bit been revealed to non-magicians about the methods of magic, even if not everything that was said provided accurate explanations for the specific feats undertaken that evening? Then again, though, was the person singled out even from the Magic Circle? Was that suggestion, too, just part of the act?

Once the doubting started, it proved difficult to halt. As I tried to recall the events of that night, I became decidedly concerned about the extent to which my memories embellished what had taken place. But then, too, I began to appreciate that my efforts to establish what had really happened were sidelining something of significance: my reconstructions and questioning afterwards were part of building a sense of the magic performed. In other words, the magic of that night was still being worked in the days afterwards (and is still so today in writing this preface). With this recognition, I resolved to take up entertainment magic.

It would be 15 years, though, before this ambition would be realized.

Still another common function of prefaces is to forward meta-instructions on how a book should be read. In the spirit of the previous account of my origin story, I do not wish to suggest *Performing Deception* offers the definitive, for-all-purposes analysis of how magic is learnt. Instead, for me, the overall aim of this book is to promote a sense of possibilities for acting in the world. Binds, conundrums and conflicting demands with the undertaking of magic are identified to promote a spirit of curiosity regarding how we meet our day-to-day experiences.

Prefaces also acknowledge limitations. In this book, I adopt a particular orientation to limits: *Performing Deception* organizes its argument around my efforts as a novice. In the roughly three-year time span covered in this book, I came to offer regular face-to-face and online magic shows, entered into professional magic societies and studied under a highly renowned artist. Through recounting my step-by-step development, I seek to attend to facets of learning that might have become forgotten or may go unappreciated by seasoned hands.

The manner in which prefaces acknowledge limitations often goes hand in hand with justifying the choices made about what was included in the book. Here, I wish to do this, too, particularly relating to how *Performing Deception* outlines some of the secret methods employed in sleight of hand and recounts the full instructions for one particular trick. Magicians are known for refraining from sharing their methods. This book details many of the reasons for this reluctance. In seeking to examine how the skills and reasoning associated with magic are learnt, the bounds of what should be disclosed in this book have been recurring concerns for me. I justify the inclusion of information on methods on two bases. First, magic societies themselves allow for the sharing of secrets in relation to research and education. The Magic Circle, for instance, permits secrets to be published in books wholly devoted to the study of magic. Perhaps more importantly, through the close investigation of a practice, *Performing Deception* is intended to invite you into becoming a student of an art form rather than only a spectator to it. Doing so requires some familiarity with its fundamentals. Still, much has been left out by way of detail. When specifics are given, they relate to beginner-level methods and tricks that provide a base-level understanding of card magic—this will enable you, as a reader, to appreciate the themes being discussed.

Still another function of a preface is to express gratitude. While *Performing Deception* recounts an individual process of development, it—like magic itself—was not something accomplished alone. My initial efforts to formulate an academic study of entertainment magic were supported by Jonathan Allen, Wally Smith and Gustav Kuhn. I wish to thank the late Harold Garfinkel in this regard as well. Not long after the 2002 *Café Scientifique* talk, I had an opportunity to meet him and get supportive feedback on my initial ideas when he visited the University of Nottingham.

The research undertaken in this book was conducted at the University of Exeter. I could hardly hope for a more supportive intellectual environment to raise questions far beyond the scope of one discipline than my department: the Department of Sociology, Philosophy and Anthropology. Among those I am indebted to are Jonathan Barry, Giorgia Ciampi, Adrian Currie, Tia DeNora, Abi Dymond, Jane Elliot, Joel Krueger, Elis Jones, Sabina Leonelli, Georgina Lewis, Simone Long, Laura Loveday, Charles Masquelier, Mike Michael, Iain Lang, Andrew Pickering, Tom Roberts, Michael Schillmeier, Emily Selove, Ric Sims,

Kirsten Walsh and Dana Wilson-Kovacs. My thanks as well to the Egenis Research Exchange Group, the Cognition and Culture Reading Group, as well as the Magic and Esotericism Group at Exeter. Most of all, my thanks to Giovanna Colombetti for years of support.

A Visiting Fellowship at Linköping University in 2018 was hugely helpful in my experimentations. My thanks in particular to Asta Cekaite, Catelijne Coopmans, Claes-Fredrik Helgesson and Steve Woolgar.

Additional thanks as well to Malcolm Ashmore, Brian Balmer, Melissa Barrett, Ann-Sophie Barwich, Kate Blackmore, Bryan Brown, Michel Durinx, Stephen Fisher, Olga Restrepo Forero, Oliver Kearns, Todd Landman, Trudi Learmouth, Eric Livingston, Susan Maret, Catherine Moorwood, Simon Pattenden, Glen Roberts, Nik Taylor, Emily Troscianko, Rachel Tyrrell, Elspeth Van Veeren, Kathleen Vogel, Susanne Weber and James Wooldridge. A number of academic groups supported the development of the ideas in this book, including SPIN (Secrecy Power and Ignorance Network), the BSA Auto/Biography Study Group as well as the University of the Third Age. The anonymous reviewers organized through Open Book Publishers also provided many useful comments.

As someone entering a long-standing performance tradition, my development has been enabled by generations of conjurors that have come before me. Today, novice and experienced magicians can turn to the internet to access training instructions in a manner unthinkable a generation ago. Of the many I have learned from online, my particular thanks to Othmarius of Othmarius Magic for his considered instructions. It has been a privilege to be a student of Dani DaOrtiz, first through a masterclass in 2019 and then through his dD School. As a mark of his commitment to sharing knowledge in magic, even before we knew each other, Dani was open to me recording his masterclass.

The Magic Circle, Magic Circle Apprenticeship Network and the Exonian Magical Society have served as nurturing groups for peer instruction and support.

The Ashburton Arts Centre provided a wonderful venue for me to develop as a public performer. My special thanks to Chris Willis and Andy Williamson for providing me with this space.

With permission, segments in the book directly draw on or present modified versions of text in Rappert, B. 2021. "'Pick a

Card, Any Card": Learning to Deceive and Conceal — with Care' *Secrecy and Society* 2(2). doi:10.1177/1468794120965367 (CC BY 4.0 license); Rappert B. 2020. 'Now You See it, Now You Don't: Methods for Perceiving Intersubjectivity' *Qualitative Research* 22(1): 93–109. doi:10.1177/1468794120965367 as well as Rappert, B. 2021. 'Conjuring Imposters' In: Steve Woolgar, Else Vogel, David Moats, and Claes-Fredrick Helgesson (Eds.) *The Imposter as Social Theory* Bristol: Bristol University Press: 147–170. Chapter 2 includes instructions for a card trick from Fulves, K. 1976. *Self-Working Cards Tricks* New York, NY: Dover reproduced under 'fair use' (or 'fair dealing') provisions. I would like to thank all of those that shared in undertaking magic with me as part of my apprenticeship into the art of conjuring. Because, after all, can there be magic without an audience? Or, for that matter, a magician?

Finally, in directing attention this way rather than that, another function of a preface is to bring to the fore some issues while deliberately sidelining others. And so is the case here.

Transcription Notes

To examine the finer details of the moment-by-moment unfolding of interactions, I employ transcription conventions derived from the field of Conversation Analysis. As *Performing Deception* is not primarily written for students of conversation, though, I only make use of a limited range of notational conventions and symbols intended to convey basic features of verbal communication. These include:

Symbol	Example
Underlying indicates points of stress.	P1: Well, attention is being <u>directed</u> BR: By who? P1: By you. Yeah, yeah.
Capital letters denote increased volume.	BR: I did it right here in the middle of the table. Was this your card? P2: REALLY. REALLY. If you like played the tape and that is what happened I won't, I would not be surprised.
Words in ((double parentheses)) are my glossing descriptions of verbal or physical action.	((Group laughter)) Not what you do here. ((spreads out an imaginary deck in his hands))
Numbers within single parentheses, such as (1.0), indicate the approximate time of the pause in seconds. The symbol '(.)' indicates a pause for a fraction of a second where one would not be expected, given the grammatical construction of the statement.	P1: I was trying to think of other examples of power (0.5) where risks are taken P3: =you don't want to be (.) disruptive.

Symbol	Example
Pairs of equal signs indicate: *(a) Either when two individuals speak with no intervening silence* *(b) When the same individual continues a statement without pause, but that statement is broken up by talk from another individual, without significant overlap in the statements.*	P2: Like you kind of, when he spins over the card you want it to be the right card= P1: =So in that P3: I know, I am like that as well, you know, I just, in fact I still don't want to know how he makes it because= P1: Yeah P3: =it's fun.
Overlapping talk between two individuals is designated by square brackets. The start of an overlap is indicated by '[' in two adjacent lines.	P4: Yes, and then this it sort of opened up the P4: [possibility BR: [Yes, yes

In addition to these conventions, symbols at the end of a statement or line are not used to indicate the end of a sentence. Instead, periods within lined numbered excerpts indicate a falling or final intonation. A question mark indicates a rising intonation. The absence of either indicates no intonational change.

In the transcribed excerpts, 'P' denotes a magic session participant's statement, and the number next to that indicates the order in which that participant first spoke in the group. For instance, 'P2' indicates the second participant that spoke in session. Any subsequent statement by this person is likewise indicated by 'P2'.

1. A Kind of Magic

How can we understand ourselves, others and the world around us? What forms of labor are entailed in doing so? How can we recognize and foster skillful ways of seeing, feeling and acting? *Performing Deception* attends to these questions by recounting the efforts associated with learning one type of performance art: entertainment magic.

In the pages that follow, I suggest possibilities that entertainment magic (a.k.a. 'modern conjuring' or 'secular magic')[1] offers us for engaging with one another. While not unique to this art form, those possibilities are primarily associated with (i) a playful orientation toward deception, and (ii) a recognition of the limits of perception.

On the first, deception is rarely held up by conjurors as their ultimate aim; however, they routinely engage in forms of action and inaction intended to mislead their audiences. At least for some, deceiving is more fundamental to this art than entertaining; while a magician might wish to amuse their audiences, they must deceive them.[2] More than this, though, conjuring as a staged activity entails mutually monitored deception between those involved. While magicians might proffer all sorts of verbal and non-verbal explanations for their feats, audiences are likely to be suspicious about how both can function as techniques of subterfuge. Magicians, in turn, craft their performances in anticipation that at least some eyes and ears are primed for tell-tale signs of chicanery. How these overall expectations meet each other—and, in doing so, make magic—is a recurring topic for this book. In contrast to many characterizations of conjuring as a one-directional exercise in control by

---

1 During, Simon. 2002. *Modern Enchantments*. London: Harvard University Press. https://doi.org/10.4159/9780674034396.
2 Comments made during Earl, Ben. 2020, July 11. *Deep Magic Seminar*. For a contrasting orientation to the rightful place of deception, see Corrigan, B. J. 2018. '"This Rough Magic I Here Abjure": Performativity, Practice and Purpose of the Bizarre', *Journal of Performance Magic*, 5(1). https://doi.org/10.5920/jpm.2018.05

magicians, however, I advance an understanding of it as a reciprocal interaction that involves the interplay of care, control and cooperation.

On the second offering of entertainment magic, learning conjuring is an unmistakably embodied endeavor. The body is a starting basis for engagements with the world, and a product of those engagements. And yet, more subtle considerations will come into play in *Performing Deception*, other than noting how an individual's trained body figures as both means and ends. I will advance conjuring as a curious art because of how the acquisition of skills invites a refinement *and* an unsettling of sensory experiences.[3] Learning magic supports recognition by the learner of how what is observable depends on our human faculties, and underscoring (again and again) that these faculties are fallible.[4] In this way, learning magic illustrates the illusionary nature of our everyday sensory ways of navigating through the world, even as our perceptions are vital to experiencing magic in the first place. I want to consider the kinds of possibilities and challenges this condition provides for rethinking our interactions with one another—if we can find ways to be receptive to the tangles of experience. Part of the intended offering of *Performing Deception* is to propose approaches, techniques and concepts for getting entangled.

Beginnings

This book adopts a particular tack in doing so. Its central spine consists of what is conventionally called a 'self-study' (or what I will come to refer to as a 'self-other' study).[5] I detail my immersion into learning conjuring as a basis for considering how practical reasoning and embodied skills are acquired.

---

3   For a cultural critique of how the body has been approached as unreliable, see Bordo, Susan. 1993. *Unbearable Weight*. London: University of California Press: Chapter 1.
4   For an example on such an undertaking, see Ekroll, Vebjørn, Sayim, Bilge and Wagemans, Johan. 2017. 'The Other Side of Magic: The Psychology of Perceiving Hidden Things', *Perspectives on Psychological Science*, 12(1): 91–106. https://doi.org/10.1177/1745691616654676.
5   This is in line with traditions across a variety of disciplines. For instance: Sudnow, D. 1978. *Ways of the Hand*. London: MIT Press; O'Connor, E. 2005. 'Embodied Knowledge', *Ethnography*, 6: 183–204; and Rouncefield, M. and Tolmie, P. (Eds) 2013. *Ethnomethodology at Work*. London: Routledge. https://doi.org/10.4324/9781315580586.

To offer an analogy, this study is not conceived as providing a panoramic view looking down on a terrain from the highest vantage point. Instead, it is an analysis of what climbing entails; it is from a starting position of comparative *ignorance* and *inability* that I want to voice appreciations of conjuring.

What can be gained by attending to the toils of a learner? The intention here is not to forward my experiences as somehow standing for every magician or beginner. Instead, my questions and quandaries serve as entry points for thinking through what is at stake in the development of competency.

For instance, looming large for me when this work began in late 2017 was an elementary matter: I had not the faintest clue what skills would be involved. At the time, my conjuring know-how was confined to a hazy memory of a couple of childhood card tricks. This ignorance was not by chance. As an activity of staging the fantastic, improbable or impossible, magic relies on covertness in method. Moreover—as mentioned in the preface—in my case the unfamiliarity was self-enforced. Since my original inspiration to undertake this research in 2002, up until late 2017 I had steadfastly avoided watching documentary-type programs on magic or even attending live performances.

Basic uncertainty about 'how magic is done' fostered a lingering doubt about my prospects for acquiring any level of proficiency. As with other arts and crafts—dancing, glass blowing or sculpting, to name but a few—through subjecting oneself to the repetitive demands of training, it is possible to cultivate new ways of acting in the world. The prospects for refining through practice, though, is mingled with basic questions about the requisite underlying abilities necessary. When I told a friend about my impending conjuring venture, she replied: 'Oh I would love to learn magic, but doesn't it take a lot of dexterity?'. I guessed that it did. As a then 45-year-old with no particular background of fine motor training, I worried about the physical demands. What was required, whether I wanted to discipline myself enough to undertake the training, as well as whether I had a sufficient level of agility to train at all, though, were matters I could only speculate about before I began.[6]

---

6 It would be many months until I appreciated that many professional magicians lament how much their colleagues rely on manufactured gimmicks which replace the need to learn physical sleights. As in comments by John Carney, see Regal, D. 2019. *Interpreting Magic*. Blue Bike Productions: 142.

It was against these kinds of uncertainties that I formulated a starting sense of how my status as a beginner would prove advantageous. As Zen monk Shunryu Suzuki advocated, the promise of a 'beginner's mind' is to be able to question what might otherwise be taken for granted or discounted.[7] As he contended, 'In the beginner's mind there are many possibilities; in the expert's mind there are few'.[8] In approaching magic without many preconceptions, I hoped to attend to aspects of this art that might be known but go unnoticed by more seasoned hands.

As this book will make clear (especially Chapter 5), experienced magicians often seek to cultivate a mind without preconceptions. This is so because they wish to understand the effects of their doings on their audiences. Being able to see with the eyes of the uninitiated, however, is by no means regarded as straightforward. A common criticism some magicians make is that their peers are too fixated on their own predilections to recognize what matters for audiences. As a result, being taken as able to speak for audiences is a prized aptitude. In no small manner, the attribution of this ability defines who speaks with authority.

These points imply that an account of magic—especially how it is learnt—should not be divorced from who is giving that account. Partial in its view, locatable within a biography and corporeally embodied, *Performing Deception* offers what Donna Haraway called situated knowledge.[9] It is an account from a shifting 'here' rather than from an abstract 'nowhere' or universal 'everywhere'.

I will characterize my learning development as aligned with the circulation between the modes educational theorist David Kolb proposed as part of 'experiential learning'.[10] Herein, learning consists of (i) taking in information through attending to lived *concrete experiences* and by using *abstract concepts and theories*, as well as (ii) transforming information through acts of *observation and reflection* and by *active experimentation* in solving problems. I take Kolb's four-part breakdown of learning not as neatly distinct categories, but instead as starting

---

7   Suzuki, Shunryu. 2005. *Zen Mind, Beginner's Mind*. London: Shambhala.
8   Ibid.: 2.
9   Haraway, Donna. 1988. 'Situated Knowledges: The Science Question in Feminism and the Privilege of Partial Perspective', *Feminist Studies* (Autumn), 14(3): 575–599.
10 Kolb, D. A. 2015. *Experiential Learning* (Second Edition). Saddle River, NJ: Pearson Education. See, as well, Ragin, Charles and Amoroso, Lisa M. 1994. *Constructing Social Research: The Unity and Diversity of Method*. London: Sage.

prompts for considering inter-relations between different ways of being in the world. I portray the processes of taking in information as well as transforming it as infused with disorientations. In my case, learning magic entailed a maturing hesitancy about what I knew, even as I became defter in physically moving objects and socially being with audiences.

To present learning as an interchange between experiences, concepts, reflections and experimentation implies that each mode needs to be understood in relation to other modes and, furthermore, that none should occupy a privileged place. Within many academic studies—including those of embodied skills training—it is commonplace to start with concepts and theories that serve as grid templates for determining what and how to analyze. In the case of studying enskillment, theories of labor and alienation, theories of gender enaction, or theories of embodiment all could serve as headline orientations. A theory-led analysis would aim to establish how the topic at hand illustrates, disproves, confounds, elaborates (and so on) this or that set of abstract concepts. Such a course is not taken in this book. Theories and concepts are not treated as externally derived reference frames that serve as the beginning and end points of study. Instead, I position them as relevant in the manner that *abstract concepts and theories* arise from and inform the medley of learning. Learning ways of reasoning and types of skill associated with performing magic meant encountering a whole array of binds, queries and uncertainties I had never anticipated at the start. Examining those knotty experiences through whatever resources seem fruitful has been my overall orientation.

To write this, though, is not to claim the topics at hand are somehow understood wholly on their terms, without recourse to starting presumptions or enduring preoccupations. At times in *Performing Deception*, its starting points will be evident. For instance, motivated early on by a desire to ask how magic provides a means for coming to understand ourselves, others and the world, in this book I frequently return to questioning the relevance of deception in its undertaking.[11] I ask how the acknowledgement that some kind of intended manipulation

---

11 For efforts to distinguish entertainment magic through reference to deception, see Luhrmann, Tanya M. 1989. 'The Magic of Secrecy', *Ethos*, 17(2): 131–165 and Villalobos, J. Guillermo, Ogundimu, Ololade O. and Davis, Deborah. 2014. 'Magic Tricks'. In: *Encyclopedia of Deception*, Timothy R. Levine (Ed.), Thousand Oaks: Sage: 636–640.

is afoot informs the actions and experiences of magicians and audiences alike. Concepts of deception, however, are positioned as relevant in relation to an overall path of learning. They inform observations, reflections and experimentation, and each of these, in turn, inform the further selection and interpretation of concepts and theories. This iterative path is how I hope to see the world through concepts but not to be (overly) blinded by them.

As elaborated in later chapters, dynamic relations between the modes of experiential learning were particularly salient for my development because my initial performance audience consisted of fellow academics. These individuals often brought to bear their own formal theories for interpreting what they had witnessed.[12] In practice, their reliance on theories aided me to selectively direct their attention through acts of simulation and dissimulation.[13]

While attending to mine and others' mix of experiences, concepts, reflections and experimentation moderates the role given to abstractions, this tack simultaneously suggests giving the nitty-gritty minutiae of our practical doings a more prominent space than they are often accorded. Such attention, in turn, is aligned with certain intellectual approaches. Harold Garfinkel characterized a central task of ethnomethodology as treating 'practical activities, practical circumstances, and practical sociological reasoning as topics of empirical study, and by paying to the most commonplace activities of daily life the attention usually accorded to extraordinary events, seek to learn about them as phenomena in their own right'.[14] So, too, is this a central task in *Performing Deception*.

---

12 For a discussion of the relevance of existing beliefs in the interpretation of magic, see Smith, W. et al. (forthcoming). *Explaining the Unexplainable: People's Response to Magical Technologies*; and Olson, J. A., Landry, M., Appourchaux, K. and Raz, A., 2016. 'Simulated Thought Insertion', *Consciousness and Cognition*, 43: 11–26. https://doi.org/0.1016/j.concog.2016.04.010.

13 Much the same can and has been said for magicians too. For an extended argument on how magicians' assumptions and concepts can lead them astray, see Ortiz, Darwin. 1994. *Strong Magic*. Washington, DC: Kaufman & Co.

14 Garfinkel, Harold. 1984. *Studies in Ethnomethodology*. Cambridge: Polity: 1. In a related but also distinctive vein, this book—at times—adopts what Tia DeNora referred to as a 'slow sociology'. See DeNora, Tia. 2014. *Making Sense of Reality*. London: Sage. https://dx.doi.org/10.4135/9781446288320. For a discussion of magic as a practical accomplishment, see Laurier, Eric. 2004. 'The Spectacular Showing: Houdini and the Wonder of Ethnomethodology', *Human Studies*, 27: 377–399. https://doi.org/10.1007/s10746-004-3341-5.

In attending to the particulars of doing, what will become plain is that learning is characterized by unruliness. As with other crafts and arts entailing bodily movement, positioning, grasping and the like, learning conjuring involves developing a direct hands-on sense of how to act. This includes how to engage with objects and material environments. However, in practice, these can rebuff our whims. In magic, everyday objects or specially ordered props can fail to bend, snap, twist, flip or slide appropriately. Material settings like a stage or a table can prove stubbornly resistant to requirements. As Andy Pickering, Annemarie Mol and others have advocated, understanding how humans act in the world requires attending to the back-and-forth dance between human and non-human agency. Each affects the other.[15] What the general acknowledgement of this dance implies is that skills development is bound to be unpredictable. And so it was for me.

Defining Magic

To preview the subsequent argument, as I have done so far in this introduction, might well be regarded as premature. After all, so far no definition has been given to the central topic under investigation. What, then, is entertainment magic?

Magic. The term routinely conjures up all sorts of associations. At times, that might be recollections of iconic acts, such as a rabbit being pulled out of a top hat. At other times, specific performers might come to mind. David Copperfield, Penn & Teller and David Blaine featured as some of the eminent individuals that formed my early impressions of this art. Magic can also be defined in terms of what it does: generating wonder, a natural state of mind, astonishment and sometimes even discomfort have been advanced as its central aims.[16] Such effects stem

---

15 Pickering, Andrew. 2017. 'In Our Place: Performance, Dualism, and Islands of Stability', *Common Knowledge*, 23(3): 381–395. https://doi.org/10.1215/0961754X-3987761 and Mol, Annemarie. 2002. *The Body Multiple: Ontology in Medical Practice*. Durham, NC: Duke University Press. https://doi.org/10.1215/9780822384151.

16 For varied statements on what effects are sought, see Sharpe, S. H. 2003. *Art and Magic*. Seattle: The Miracle Factory; Brown, Derren. 2003. *Absolute Magic: A Model for Powerful Close-Up Performance* (Second edition). London: H&R Magic Books. Harris, Paul and Mead, Eric. *The Art of Astonishing*. [n.p.]: Multimedia A-1; Fitzkee, Dariel. *Magic by Misdirection*; Burger, E., and Neale, R. E. 2009. *Magic and Meaning*. Seattle: Hermetic Press.

from the 'juxtaposition between the conviction that something cannot happen and the observation that it just happened.'[17] Instead of merely challenging audiences to discover how effects are produced, some professionals take their job to be one of creating illusions that truly convince audiences that the impossible has been made possible[18]…at least for a short time.[19] For some, 'magic' as a designation should be reserved only for displays of the impossible in which the magician has calculatingly cancelled out every reasonable explanation audiences might harbor.[20]

Alternatively, entertainment magic can be recognized through its kindred affiliations. Even in modern times, that company has varied widely. Reading the mind of a member of the audience has been aligned with paranormal abilities, mystical energies and much more besides. Just as magic can be defined through its affiliations, it can be defined through its *disaffiliations*. Reading the mind of a member of the audience can be overtly presented as decidedly *not* based on paranormal abilities, mystical energies and much more besides.

The diverse pictures and feelings magic summons point to the scope for disagreement about just what ought to be conjured up through evoking this term.[21]

Consequently, in examining entertainment magic, a trick I will need to execute throughout this book is how to both trade on settled notions of what magic entails (to build on others' prior observations), while also calling into doubt settled notions (to question certain presumptions). The need to do so for this specific term is just one instance of many in which I will both marshal and question prevalent concepts, conventions and categories.

In acknowledging this tension, I want to offer the characterization of entertainment magic as *deft contrariwise performance*. The purpose

---

17 Lamont, Peter. 2009. 'Magic and the Willing Suspension of Disbelief'. In *Magic Show*, Jonathan Allen and Sally O'Reilly (Eds). London: Hayward Publishing: 30.
18 For instance: see Aronson, Simon. 1990. *The Illusion of Impossibility*. [n.p.]: Simon Aronson: 172; Tamariz, Juan. 2019. *The Magic Rainbow*. Rancho Cordova, CA: Penguin Magic; and Olewitz, Chloe. 2020. 'Francis Menotti's Weird Words', *Genii* (November): 39.
19 Nelms, Henning. [1969] 2000. *Magic and Showmanship*. Mineola, NY: Dover.
20 Leddington, Jason. 2016. 'The Experience of Magic', *The Journal of Aesthetics and Art Criticism*, 74(3): 253–264.
21 Allen, Jonathan. 2007. 'Deceptionists at War', *Cabinet* (Summer), 26. http://www.cabinetmagazine.org/issues/26/allen.php

of doing so is not to assert a universal definition that sorts out once and for all what really counts as conjuring 'proper'. Instead of sticking labels, my purpose is to advance a set of sensitivities for considering the possibilities magic provides for understanding ourselves, others and the world.

Let me expand. *Deft* speaks to how proficiency in conjuring requires adroitness and artfulness. The want for such skillfulness, though, is being conceived in a particular way that might counter some readers' expectations. Deft could be taken to apply to the appearances of handling techniques. In this vein, a deft performer is someone who smoothly, confidently and neatly manipulates the apparatus of magic—cards, coins, balls and the like.[22] Or, deft might refer to someone with a silver tongue who confidently commands a floor. While magicians often seek to act in ways regarded as nimble, flashy or adroit, often times they do not. Appearing chaotic, clumsy or even out of control in the eyes of audiences is one way of hiding jiggery-pokery.[23] Struggling can help excite drama too.[24] Thus, magicians can be regarded as virtuoso performers without displaying virtuosity in their movements.[25] Deft, then, does not signal a finished quality of appearance, but rather an orientation to appearances. Appreciating when and how to display manual or other abilities is part of the competency developed in learning magic.

As implied by the previous paragraph, conjuring needs to be understood through reference to both the actions and experiences of all those taking part in it. In this vein, to characterize magic as a type of *performance* is to signal how it entails individuals engaging with each other (physically or remotely) through practices, conventions and rituals. The *performance* in *deft contrariwise performance* is understood in an expansive manner. Following sociologist Erving Goffman, performance is conceived here as 'all the activity of a given participant

---

22 For a statement along these lines see Garcia, Frank. 1972. *Million Dollar Cardsecrets*. New York: Million Dollar Productions.
23 For practitioner discussion of this, see Youell, Steven. 2009. *Weapons of Mass Deception*. Lecture notes: 45–47. For an analysis on how being out of control is marshalled in performance, see Jones, Graham and Shweder, Lauren. 2003. 'The Performance of Illusion and Illusionary Performatives: Learning the Language of Theatrical Magic', *Journal of Linguistic Anthropology*, 13(1):51–70.
24 Clifford, Peter. 2020. *A Story for Performance*. Lecture notes from presentation at The Session (London), 12 January.
25 As a result of the previous points, what counts as a virtuosic performance is as up for dispute as what makes magic 'magical'.

on a given occasion which serves to influence in any way any of the other participants'.²⁶

Three points of clarification to this definition of *performance* need to be made at this stage. First, magic in this book is not simply conceived in relation to the activities of one figure—the magician. While a conjuror might be the center of attention and might be called 'the performer', audience members are approached as meaningful agents. A goal of this argument is to describe how. Second, magic is approached as a coupling of people and material things. As such, the material world can be conceived of as a meaningful agent in what unfolds. Again, a goal of this argument is to describe how. Third, entertainment magic has a complex relation to the recognition that a performance is taking place. At one extreme, overtly billed instances of conjuring, like a Las Vegas stage show, might be widely appreciated as contrived performances. Yet it is just this recognition that hazards those present dismissing or downplaying what they witness as 'mere' trickery. At the other extreme, efforts to weave displays of the impossible into everyday settings without the conventional trappings of 'a magic show' risk being dismissed or downplayed as mere coincidence, mistaken perception and so on. As a result, how a scene is defined by those involved is of no small significance to how and whether the label of 'magic' applies.

Whilst not an everyday or scholarly word, the various definitions of *contrariwise* speak to important considerations in conjuring. One definition of contrariwise is 'contrary to expectations'.²⁷ Magic often entails spectators observing outcomes the audience believes could not have taken place.²⁸ How can coins be plucked out of mid-air—one after another, and another and another? Just what is considered contrary to expectations, though, is not fixed. Conjuring exists in a dynamic relation to cultural beliefs because it seeks to defy some of them and, in doing so, helps redefine what counts as a valid belief. Everyday notions of what is possible have shifted over time, not least because of technology. As a result, conjurors have adapted their routines to cultural expectations in

---

26 Goffman, Erving. 1956. *The Presentation of Self in Everyday Life*. New York: Doubleday: 8.
27 *Memidex*. http://www.memidex.com/contrariwise+to-the-contrary
28 Lamont, Peter. 2013. *Extraordinary Beliefs*. Cambridge: Cambridge University Press. https://doi.org/10.1017/CBO9781139094320.

order to keep their methods obscured.[29] But more than this, conjurors have marshalled the commonplace beliefs of their day about science, technology and magic itself to misdirect. The last of these is perhaps particularly noteworthy. The magician turned espionage consultant John Mulholland spoke to this point in his secret manual for the US Central Intelligence Agency. For him, defying others' presumptions about how deception is achieved was integral to successful trickery.[30] Entertainment magic is accomplished—and covert espionage operations as well—when the performing actors play off the beliefs of their audiences (whether erroneous or valid).

Another definition of *contrariwise* is 'from a contrasting point of view'.[31] Magicians often seek to foster an impression in audiences at odds with their own understanding. Shuffling cards might be taken by onlookers as a process of disordering. Yet, for those doing the manipulations, shuffling can be a process of ordering. As such, practicing magic routinely requires trying to imagine others' perception of what is on display rather than relying on what one knows to be the case. Darwin Ortiz spoke to this point in contending that the impressiveness of an effect *'depends on your audience's perceptions, not on the reality of the situation'.*[32] When magicians perform for audiences with varying familiarization with magic, approaching the performance from contrasting points of view is vital. This is so because what generates awe in those with no knowledge of the methods of magic may not do so for old hands. When magicians perform across cultures, the demands on their art can be considerable.[33]

Contrariwise also means 'in a perverse manner'.[34] Certainly, magic can be performed in troublesome ways. For instance, this happens when

---

29 Bell, Karl. 2012. *The Magical Imagination.* Cambridge: Cambridge University Press; Mangan, Michael. 2007. *Performing Dark Arts: A Cultural History of Conjuring.* Bristol: Intellect; and Smith, Wally. 2015. 'Technologies of Stage Magic: Simulation and Dissimulation', *Social Studies of Science*, 45(3):319–343. https://doi.org/10.1177/0306312715577461
30 Mulholland, John. 2010. In *The Official CIA Manual of Trickery and Deception*, H. Keith Melton and Robert Wallace (Eds). London: Hardie Grant: 69–81.
31 *American Heritage® Dictionary of the English Language* (Fifth Edition). 2011. https://www.thefreedictionary.com/contrariwise
32 Ortiz, Darwin. 1994. *Strong Magic.* Washington, DC: Kaufman & Co.: 70 (italics in original).
33 Palshikar, Shreeyash. 2007. 'Protean Fakir', *Cabinet* (Summer). http://www.cabinetmagazine.org/issues/26/
34 *American Heritage® Dictionary of the English Language* (Fifth Edition). 2011. https://www.thefreedictionary.com/contrariwise

magicians take themselves too seriously, or when they fail to take the magic seriously.[35] On the former, with asymmetries between performers and audiences regarding who speaks and who directs, the potential entertainment value of magic exists alongside its potential for inflicting humiliation and domination.[36] In short, it is a double-edged activity. The double edge, in part, derives from how establishing a human(e) connection with others is both frustrated and underpinned by secrecy and deception. The potential inflicting domination also stems from how magic enacts wider cultural beliefs. As such, what counts as entertaining can reinforce questionable stereotypes of the time on matters of gender, class, race and much else.[37]

At perhaps its most general definition, *contrariwise* can simply mean 'in the opposite way'.[38] In my experience, learning magic requires cultivating opposing manners of reasoning. An example of this is the ability to shift between different orientations to sensory experience. On the one hand, as with so many other arts, learning consists of developing something of an 'eye'.[39] A learner begins to appreciate and harness visual and other sensory subtleties that would pass by the uninitiated. Concerning his apprenticeship as an amateur boxer, for instance, sociologist Loïc Wacquant described how he acquired the 'eye of a boxer' that enabled him to pick up on otherwise invisible movements.[40] I, too, developed newfound appreciations.

Conversely, because of how magic often utilizes the bounds of our cognitive and perceptual capabilities, learning entails becoming (newly) aware of the limits of what we can perceive. Thus it requires a double movement: closely attending to, and coming to doubt, sensory

---

35 Burger, Eugene and Neale, Robert. 2009. *Magic and Meaning*. Seattle: Hermetic Press.
36 During, Simon. 2002. *Modern Enchantments*. London: Harvard University Press: 131–132. https://doi.org/10.2307/4488584
37 Goto-Jones, Chris. 2016. *Conjuring Asia: Magic, Orientalism, and the Making of the Modern World*. Cambridge: Cambridge University Press.
38 *American Heritage® Dictionary of the English Language* (Fifth Edition). 2011. https://www.thefreedictionary.com/contrariwise
39 See, for instance, O'Connor, E. 2005. 'Embodied Knowledge', *Ethnography*, 6, 183–204; Ameel, Lieven and Tani, Sirpa. 2012. 'Everyday Aesthetics in Action: Parkour Eyes and the Beauty of Concrete Walls', *Emotion, Space and Society*, 5: 164–173; and Roepstorff, A. 2007. 'Navigating the Brainscape'. In: *Skilled Visions: Between Apprenticeship and Standards*, C. Grasseni (Ed.). Oxford: Berghahn Books: 191–206.
40 Wacquant, Loïc J. D. 2004. *Body & Soul: Notebooks of an Apprentice Boxer*. Oxford: Oxford University Press: 117.

experiences. As a result, one does magic, but the magic does something back.

Acting 'in the opposite way' also gestures toward another facet of magic implicit in the previous points: the scope for contrasting recommendations about how it should be done.[41] For instance, a long-standing principle is to never foreshadow what is about to be done.[42] To do so provides the audience with information that may enable them to figure out how a feat was accomplished (or, just as bad, to believe they have figured it out). And yet, for one of the towering figures in magic today, Juan Tamariz, skillfully previewing the effect to be performed can add much to the feelings of astonishment generated.[43] A student of Tamariz, Dani DaOrtiz, has gone further to advocate that foreshadowing what is about to happen should be integral to performances. In doing so, conjurors can powerfully affect the expectations, and thereby the emotions, of spectators.[44]

In general, as a continuously developing art form, much of the innovation in conjuring derives from attempts to depart from previously established conventions.[45] At the level of individuals, it is through offering a distinctive presentation style that magicians develop as artists.[46] As an example, a conventional way of characterizing magic is as a performance that draws 'the audience towards the effect and away from the method'.[47] While this is often the case in the kind of magic under study in *Performing Deception*, it is not always so. Performers such as Penn & Teller have used the selective revelation of methods to affect their audiences.

The term *contrariwise*, then, speaks to many facets of magic.

---

41 In Western history, the term magic has been repeatedly applied to those activities deemed in opposition to conventional and acceptable forms of belief and practice; see Rally, Robert. 2010. *Magic*. Oxford: Oneworld.
42 See Jones, Graham. 2011. *Trade of the Tricks*. London: University of California Press: 52.
43 Tamariz, Juan. 2019. *The Magic Rainbow*. Rancho Cordova, CA: Penguin Magic: 166–173.
44 DaOrtiz, Dani. 2018. *Working at Home*. Grupokaps: 133.
45 Allen, Jon. 2013. *Connection*. Las Vegas, NV: Penguin Magic.
46 Regal, David. 2021, February 9. *Bristol Society of Magic Lecture*.
47 Lamont, Peter and Wiseman, Richard. 1999. *Magic in Theory*. Hatfield: University of Hertfordshire Press: 31.

'Both-And' Relations

Further than just characterizing magic as a back-and-forth of this-and-that, I want to advance a specific way of conceiving the relation between elements. Inspired by Communication Studies scholars Leslie Baxter and Barbara Montgomery's dialectical approach to interpersonal communication,[48] the remaining chapters structure the analysis of learning and performing magic through attending to how they entail opposing but yet co-existing tendencies and features. For instance, as Baxter and Montgomery note, a common fault line in personal relations is how parties negotiate connection and separation. Rather than treating them as opposite states in which couples are either independent or interdependent, Baxter and Montgomery ask how relations invariably involve an interplay between both such tendencies.[49] For instance, achieving a connection is dependent on a sense of there being distinct identities in the first place. Likewise, a sense of individual autonomy is realized through one's relations to others. As a result, to place couples along a spectrum of separation-connection obscures much of the subtlety of personal relations.

It follows from these points that the presence of notionally opposing features is not something to be avoided because it is necessarily disharmonious. Instead, pushes and pulls this way and that are often inescapable. What matters, to draw on the words of the philosopher John Dewey, is the way tendencies 'bear upon one another, their clashes and unitings, the way they fulfil and frustrate, promote and retard, excite and inhibit one another.'[50] Knowing how to act is messy and subject to revision. When considering intimate personal relations, what blend of independence and interdependence is fitting at a given moment depends on the specific history of a couple, as well as how that relationship aligns with wider societal expectations.[51]

---

48 Baxter, Leslie A. and Montgomery, Barbara M. 1996. *Relating: Dialogues and Dialectics*. London: Guilford.
49 For a further extension of Baxter and Montgomery's approach to face-to-face interactions, see Arundale, Robert B. 2010. 'Constituting Face in Conversation: Face, Facework, and Interactional Achievement', *Journal of Pragmatics*, 42: 2078–2105. https://doi.org/10.1016/j.pragma.2009.12.021.
50 Dewey, John. 1934. *Art as Experience*. New York: Perigee Books: 134.
51 Furthermore—as a higher possibility—the qualities of freedom and captivity can be understood as inter-related rather than mutually incompatible and discordant.

In regarding magic as a *deft contrariwise performance*, I adopt a dialectical orientation to escape from delimited conceptualizations of skill, concealment, control and other notions. With this orientation, the ability to creatively work with contrary tendencies is part of what distinguishes the adroit from the not. This 'both-and' orientation exemplifies the spirit of curiosity that has been central to my personal development as a performer. As later chapters detail, my learning did not just consist of gaining new skills and forms of reasoning, but also a sense of the fraught conditions for learning. I came to know, to realize I did not know, to wonder what I could know, and to doubt what I thought I knew through my engagements with others. Stated in different terms, knowledge and ignorance were both mutually constitutive of learning. This 'both-and' orientation is also justified because it provides a basis for acknowledging the contests between conjurors regarding what conduct is appropriate, impactful, meaningful, etc. Indeed, more than just acknowledging the presence of contests, a dialectical orientation suggests the advisability of fostering contests if art forms are to avoid stagnation.

In short, in approaching conjuring as *deft contrariwise performance*, I intend to signal how the undertaking of magic can be understood as dynamic interplay; that is, as a relation of varied considerations that are taken to complement and oppose each other. Again and again in the pages that follow, enculturation into magic will be presented as learning how to position ways of doing, thinking and feeling that are co-existing and conflicting. A central aim of the chapters is to characterize the dynamics whereby self-other, control-cooperation, truth-deception and so on co-exist and conflict.

Chapters

As an art based on esoteric information and embodied know-how, with little in the way of established instructional institutions or accreditation procedures, how can new entrants to magic develop? Each chapter attends to activities designed to promote learning: instructional texts and videos, training demonstrations, scientific articles, recorded shows and autobiographies. In engaging with such material, each chapter addresses seemingly contrary tendencies identified within conjuring to assess how they are said to inter-relate, and then to consider the options for how they could be realized together.

Along these lines, Chapter 2 takes as its focus the relationship between *self* and *other*. This is a complex interweaving; as individuals, we cannot be understood as existing completely separate from others, and yet others are clearly not the same as ourselves. How, then, to characterize the relationship between people? I consider this in the case of magic by beginning where my study began: reading written instructions for novice card tricks. Aligned with studies in the field of ethnomethodology, in this chapter I attend to the varied forms of work associated with enacting written instructions. Prominent among them was trying to experience the magic as an audience member. Although making sense of the instructions was a solitary activity (in the sense of being undertaken alone) it repeatedly entailed imagining how tricks would affect others. This imagining was tension-ridden, not least because becoming familiar with the methods for magic had the result of changing my appreciations of what shuffling, picking and naming cards can occasion. The dance between being able to connect with others and becoming estranged from them serves as a central topic of this chapter and a recurring theme for the book.

Chapter 3 offers an understanding of magic as a form of group interaction. I recount my initial experiences in performing magic for audiences, and in particular how we produced and coordinated our conduct in ways that blended *cooperation* and *control*. Based on an innovative group method, I offer a non-conventional view of the performance of conjuring. It is non-conventional in the manner it seeks to de-center magicians. It does so by moving away from conceiving of magic as a performance by conjurors who render their audiences into manageable objects. Instead, I advance the notion of 'reciprocal action' to signal how magic can be understood as an interaction. Herein, audiences can play an active role in enabling deception and concealment, both through how they go along with and how they contest the directives of conjurors.

Chapters 2 and 3 attend to dynamics of intersubjectivity—for instance, how learning and performing magic involves both using analogic reasoning to comprehend how others experience the world, while also appreciating how others can have dissimilar experiences that are out of reach. In doing so, these chapters establish a central tension in this book: how individuals can simultaneously be brought together and disconnected by deception.

Our undertakings in the world are not simply person-to-person. Instead, they are materially mediated and enabled. Chapter 4 considers how people and the material world are coupled together in conjuring. It does so through the central notion of *naturalness*. Learning magic entails disciplining one's movement and comportment. It also entails cultivating an awareness of how material objects and settings are positioned. One central goal for doing both in modern styles of magic is to make performances look natural, spontaneous and ordinary—and thus expected, justified and above board. And yet, achieving this appearance within a conjuring performance is widely regarded as a hard-won accomplishment involving highly *contrived* actions undertaken with potentially unruly objects and others. This chapter addresses several key questions: how is naturality made intelligible as a feature of action? And how is naturality accomplished specifically within the manufactured setting of a magic show? How are learners taught to appear natural? I map the varied responses given to these questions through reviewing the arguments of prominent professionals as part of written, audiovisual, and face-to-face forms of instruction.

Chapter 5 turns to the interplay of *proficiency* and *inability*. For many types of physical crafts, the mass of manual skills involved are difficult to recount by practitioners because they have become implicit.[52] In the case of magic, the relation between what is on display and the proficiency of the performer is difficult to establish for audiences because the methods at work are obscured. Thus, what is captivating to one audience might not require sophisticated physical skills. Conversely, physically and mentally demanding feats may generate little notice. A further complication in relation to what is displayed to the underlying skills required is that learning magic entails coming into an appreciation of the limits of human perception.

To explore these issues, Chapter 5 attends to the coupled matters of how perception underlies claims to proficiency, as well as how perception is accomplished in specific situations. It begins by outlining some of the

---

52 Suchman, Lucy. 1987. *Plans and Situated Action*. Cambridge: Cambridge University Press.

varied ways magicians give place to technical ability and expert authority. Next, I turn to my performance experiences as a novice, with particular emphasis on how notions of my skill were made relevant within the moment-to-moment unfolding of interactions. In doing so, I elaborate further on points made in previous chapters regarding how magic as an activity is constituted by audiences. Following on from the initial sections, this chapter considers how expertise, proficiency and authority are enacted within instructional settings. I do so through detailing face-to-face training offered as part of a masterclass I undertook with the world-renowned magician Dani DaOrtiz. Part of this analysis includes consideration of how his teaching called into doubt the reliability of students' bodily senses and common forms of reasoning, even as our senses and reasoning as students provided the basis for validating his instructions. Through its varied components, this chapter assesses how appeals to perceptions are used to evidence, demonstrate and challenge notions of who is able to appreciate what is right in front of them.

Chapter 6 turns towards the place of a specific kind of skill cultivated in magic: the ability to sincerely deceive. It does so through examining a particular type of writing which is significant for those seeking to know about the ins and outs of this art: autobiographies. Autobiographies serve as an interesting source for investigation because of how they handle competing demands. On the one hand, this genre is typically built on appeals to authenticity; writers offer readers a backstage view of their lives, experiences and inner thinking. In doing so, this genre generally calls for a demonstration of sincerity. In contrast, much of the aura associated with magicians lies in their ability to dissemble. How then do conjuror-authors fashion their life stories such that they can hold together evidence of their genuineness with evidence of their ability to deceive? How are *truth* and *deception* positioned as part of their accounts? Chapter 6 takes up these and related questions through orientating to autobiographies as 'no less theatrical than other performances'.[53]

Not least because of the highs and lows of the hours and hours spent refining minute hand movements, the process of learning magic can be accompanied by a recurring question: why do this? This question is often accompanied by another one: how? Especially because of the presence

---

53 Allen, Jonathan and O'Reilly, Sally. 2009. *Magic Show*. London: Hayward Publishing: 46.

of covertness and the asymmetries in action between magicians and audiences, conjuring is often recognized by its practitioners as a fraught moral activity. In Chapter 7 I want to draw out mutual dependencies in magic by approaching it as an interplay of *care* and *control*. As with other chapters, the starting orientation is not to treat caring and controlling as opposites. Instead, drawing on feminist and other theories of care, I treat caring as entailing forms of controlling, and controlling as enabling forms of caring.

As a way of suggesting the possibilities for being and doing in the world, Chapter 7 also outlines the overall rationale and structure that emerged for my public performances. Instead of approaching magic as a means of accomplishing extraordinary feats with ordinary objects, I framed my performances in terms of appreciating the ordinary. The ordinary here referred to our day-to-day interactions—how we manage to live together with one another. Rather than effectively sweeping audiences away, my goal became one of finding ways to bring them back to the wonder of how we interact together; to the alluring power that is invested in secrets; to how we make perceptual sense of the world with one another and so on. This was done through offering tricks, verbal patter and questioning that took as their topics our very interactions together there and then. In this sense, my style aligned with what Augusto Corrieri referred to as 'meta-theatre'; that is, a form of action that promotes 'self-reflexive interrogation of the status of the act itself'.[54] In my case, though, the self-reflexivity manifested itself in a collective discussion about the interactional dynamics that make magic as part of the making of magic. In this way, an objective was not to have spectators to a show, but participants to a dialogue.

The final chapter offers some concluding points by returning to a question at stake throughout the book: what is magic? In closing, I seek to foster a spirit of curiosity, suppleness and questioning that helps enable novel ways of doing and being. This is done, in part, by revisiting the meaning of other terms central to *Performing Deception*: learning, self, other, method and skill.

---

54   Corrieri, Augusto. 2018. 'What Is This...', *Platform*, 12(2): 16.

2. Self and Other

Who we are as individuals depends in no small part on our relations with others.

The interplay between connection and separation has figured centrally in many attempts to theorize human relations.[1] Along these lines, families can be thought of as constituted through how their members mix interdependence and independence, as well as unity and difference.[2]

In *Performing Deception*, I approach magic as a kind of method for understanding ourselves and others. Herein, self and other are not discrete, pre-existing objects that can be plucked out of a top hat with a cry of 'Ta-da!' Instead, they form and dissolve as part of ongoing engagements. As the beginning of a much larger story about the relations between magicians and audiences, this chapter concentrates on my initial forays into learning. Through recounting the mixture of experiences, concepts, reflections and experimentation associated with practicing my first trick, I want to characterize some of the conspicuous and subtle types of work associated with magic as a domain of reasoning and skill. In particular, I attend to how notions of self and other are implicated in undertaking magic.

A Self(-Other) Stdy

But first, some basics. In seeking to understand aspects of the world, social inquiry often takes the form of an immersion into what is, at

---

1 Bakhtin, Mikhail. 1981. *The Dialogic Imagination*. Austin, TX: University of Texas Press.
2 Baxter, Leslie A. and Montgomery, Barbara M. 1996. *Relating: Dialogues and Dialectics*. London: Guilford; Arundale, Robert. 2010. 'Constituting Face in Conversation', *Journal of Pragmatics*, 42: 2078–2105.

least for the investigator, unfamiliar terrain. In turning toward the learning of embodied skills, the topic under investigation becomes one of how individuals hone ways of seeing, feeling, thinking and acting.[3] So-called 'self-studies' of acquiring practical knowledge and embodied skills involve a researcher using their own experiences of becoming a competent salsa dancer, clay sculptor, jiu-jitsu fighter and so on as a way into appreciating what a pursuit entails.

Although hardly unique to self-studies, the question of how to relate one particular pathway to others is highly salient.[4] The one-many relation, in part, turns on the status accorded to personal experience. Camilla Damkjaer spoke to this point when she contended that: 'What is important is not my subjective experience as such, but the questions and difficulties that I encounter and what they can tell me about the art of circus performance, and the possibilities created by physical reflection for an academic researcher'.[5] For Damkjaer, first-person accounts were not granted a privileged status, but they were taken as vital for knowing about the lived experiences of what it is like to perform, in her case, on a vertical rope.

In broad terms, *Performing Deception* adopts a similar set of starting premises. However, just as magic will be interpreted as entailing a shifting interplay between ostensibly opposed tendencies (see Chapter 1), so too will the study of it. In this spirit, I treat the issue of how to relate the one to the many as a matter to be revisited throughout this book, rather than as something to be set out at the start.

Also, in *Performing Deception* I orientate to magic as a thoroughly relational undertaking. While playing the piano or juggling balls can be done solo or in the company of others, it makes little sense to speak of performing magic alone. As with teachers and students, as well as joke-tellers and listeners, magicians and audiences realize themselves in relation to one another. It is this interdependency that means learning magic is poorly conceived as a self-study. Instead, it is also a study of the

---

3   For instance, see Sudnow, D. 1978. *Ways of the Hand*. London: MIT Press; as well as Tolmie, Peter and Mark, Rouncefield. 2013. *Ethnomethodology at Work*. London: Routledge.
4   O'Connor, E. 2005. 'Embodied Knowledge', *Ethnography*, 6: 183–204. https://doi.org/10.1177/1466138105057551 and Atkinson, P. 2013. 'Blowing Hot', *Qualitative Inquiry*, 19(5): 397–404. https://doi.org/10.1177/1077800413479567.
5   Damkjaer, Camilla. 2016. *Homemade Academic Circus*. Winchester: iff: 39.

possibility of apprehending others. For this reason, I refer to this book as a 'self-other study'.

Perhaps most distinctly, *Performing Deception* adopts a complex orientation to the status of personal experience. As exemplified later in this chapter, one advantage of self-studies of skill acquisition is that they make available for examination an array of embodied sensory experiences through conscious introspection. Such phenomenal experiences would be difficult, if not simply downright impractical, to access in others through techniques such as interviews or surveys. And yet, introspection, to the extent it could even be considered a method, is hardly regarded as unfailing. Beyond the commonplace kinds of doubts that might be voiced about our ability to know and describe our own experiences, this study into learning magic provides additional ones. This is so because witnessing magic—again and again—makes it clear that our senses and ordinary ways of understanding are fallible.

Therefore this 'self-other study' not only tries to unpack a phenomenon but also unpacks how that phenomenon comes into understanding. The attention to what is known and the means of knowing creates both challenges and opportunities. To discuss such points now, though, is perhaps to get ahead of the argument...

Beginnings

How can a self-other study be begun? The question has particular significance for entertainment magic due to the comparative absence of conventional pathways for training. Many other types of performance art are enculturated through professionally sanctioned programs, offered as part of established educational settings such as universities, schools and studios by accredited practitioners. Through processes of immersion, these programs have as their task preparing new entrants into a 'community of practice'.[6]

Such formal training programs, though, are relatively rare in the case of magic. Local clubs and professional societies can provide important collective settings for being with others by exchanging skills, testing competencies and developing a sense of shared identity.[7] However, their

---

6   Wenger, E. 1999. *Communities of Practice: Learning, Meaning, and Identity*. Cambridge: Cambridge University Press.
7   Jones, Graham. 2011. *Trade of the Tricks*. London: University of California Press.

availability and make-up vary widely. Today, in an era of mass online tutorials and forums, magic societies play less of a vital role than they did previously in providing access to coveted techniques. In addition, participation in a club or society is not a requirement for professionally working in the UK or many other countries.

In short, informal pathways for training are typical.[8] The comparative absence of formal training and accreditation procedures has significant implications for the development of skill, the formation of identity as well as the governance of community norms. These matters will be explored in later chapters. In late 2017 when I began practicing, I did not have a sense of such wider issues. Instead, as a novice, I was faced with a basic question: what now?

Based on a suggestion from the academic-magician Wally Smith, my pathway began with a resource central to many aspirants in the past: instructional books. Against the patchy availability of face-to-face instructions, specialized instructional books have proven a prime means of reconciling the competing desires in conjuring to delimit access to the information about the hidden methods, to enable new entrants into this art by sharing information, as well as to recognize (and reward) the contributions of innovators.

As part of its extensive magic collection, Dover Publications published eleven 'self-working' books by Karl Fulves. First printed in 1976, *Self-Working Card Tricks: 72 Foolproof Card Miracles for the Amateur Magician* initiated this Dover series, and this volume is where I began. While no definition of 'self-working' is given within the book, Fulves describes the tricks set out as 'easy to master' because they require 'no skill'.[9]

My starting orientation differed. It was, instead, informed by the long-running distinction in social research between concrete actions and their description. As one aspect of the overall distinction, scholars across diverse academic disciplines have considered the work needed to move from formalized instructions to situated action.[10] Effort is required

---

8 Rissanen, O., Pitkänen, P., Juvonen, A., Kuhn, G., and Hakkarainen, K. 2014. 'Professional Expertise in Magic—Reflecting on Professional Expertise in Magic', *Frontiers in Psychology*. https://doi.org/10.3389/fpsyg.2014.01484
9 Fulves, K. 1976. *Self-Working Card Tricks*. New York: Dover: v.
10 Suchman, Lucy. 1987. *Plans and Situated Action*. Cambridge: Cambridge University Press and Garfinkel, Harold. 2002. *Ethnomethodology's Program*. Oxford: Rowman and Littlefield: Chapter 6.

because instructions are abstractions that cannot anticipate all possible relevant contingencies. They are incomplete. As a result, readers must manage the relevance of instructions, what it means to adhere or deviate from them, what consequences are likely to follow from action, and so on. In this sense, instructions do not function so much as standards that dictate what should be done, but as resources for undertaking situated action whose meaning is settled in undertaking the action. And yet, despite what might be taken as their limitations as abstractions, instructions often serve as adequate resources for satisfactorily accomplishing tasks—assembling a cabinet, preparing a meal or fixing a leaky faucet.

Through engaging in wide-ranging forms of practical reasoning— from how to play checkers, to how to construct origami figures, to how to follow a laboratory chemistry manual—Eric Livingston concluded that: 'Realizing what [...] instructions describe depends on the work that we do to find their adequacy. The ability to find their adequacy is, to some extent, what "skill" is.'[11] Therefore, in learning conjuring, I took the gross and subtle efforts undertaken in enacting instructions as my topic for reflection and observation when I opened *Self-Working Card Tricks* on an already dark winter afternoon in late 2017.

Attending to how practical activities are accomplished is no straightforward task. Among other challenges, doing so requires contending with what Garfinkel called the 'holy hellish concreteness of things'.[12] This expression points toward the endless volume of detail that can be relevant when experience is taken as the topic of inquiry.

The next section examines instructions for a single card trick with a view to considering how notions of self and other can be implicated in interpreting texts.

As a way into, rather than out of, holy hellish concreteness, I would strongly recommend you put this book down and find a deck of playing cards to practice the instructions for yourself. Whether a new or old hand to card magic, attending to how instructions are fashioned will likely greatly enhance what you take away from your time spent with

---

11   Livingston, Eric. 2008. *Ethnographies of Reason*. London: Routledge: 100. https://doi.org/10.4324/9781315580555
12   Quoted from Liberman, Kenneth. 2007. *Dialectical Practice in Tibetan Philosophical Culture*. London: Rowman & Littlefield: 37.

this chapter. As conjuring is a bodily undertaking, there is no substitute for a bit of DIY. Expending effort in this way is also highly economical. It will raise for you subtleties that simply cannot be elaborated here—no matter your patience. Or mine.

Enacting the instructions will also aid in appraising the abstracted account of my experiences given below.[13] This account is not intended as a universal reading of the instructions. Instead, it is offered as a particular instance of sense-making, one that is of interest for how it is both the same and different from other readings. This being so, contrasting your experience based on your own personal knowledge, intentions and so on with my account provides a rare prospect in social analysis. This is a chance for you to encounter the phenomenon being analyzed akin to how the author encountered it. This is an opportunity not to be missed.

Practical Skills and *No-Clue Discovery*

Box 1 provides the instructions for the first entry in *Self-Working Card Tricks*, an entry titled *No-Clue Discovery*. It is an example of card magic that uses a Key Card Principle, a principle whose recorded origins date back to at least the 19th century.[14] Added paragraph numbers are provided for ease of reference. The photographs approximate the original sketches.

Box 1: No-Clue Discovery

1. A spectator chooses a card and returns it to the deck. He then cuts the deck and completes the cut. His card is lost in the pack and no one—not even the magician—knows where the card is.

2. The magician takes the deck and begins dealing cards one at a time into the face-up heap on the table. As the magician

---

13 For results of an audience experiment that employs the central elements of this trick, see Smith, W. et al. (forthcoming). *Explaining the Unexplainable: People's Response to Magical Technologies*.
14 More specifically, Professor Hoffman. 1876. *Modern Magic*. Eastford, CT: Martino Fine Books. See https://www.conjuringarchive.com/list/category/960.

deals, he instructs the spectator to call out the names of the cards. The spectator is asked to give no clue when his selected card shows up. He is not to pause, hesitate, blink or change his facial expression. Nevertheless, the magician claims, he will be able to detect the faintest change in the spectator's tone of voice at the exact instant the chosen card shows up.

3. The cards are dealt one at a time off the top of the deck. The spectator calls them out as they as are dealt. It does not matter how he calls them out; he can disguise his voice, whisper, shout or name the cards in French; when the chosen card turns up, the magician immediately announces that it is the card selected by the spectator.

4. *Method*: This trick makes use of a principle known as the Key Card. Before performing the trick, secretly glimpse the bottom card of the deck. This can be done as the deck is being removed from the card case. In Figure 1, the Key Card is the 3D*.

Fig. 1

5. Hold the deck face-down in the left hand. Then spread the cards from left to right, inviting the spectator to choose a card from the center, as in Figure 2.

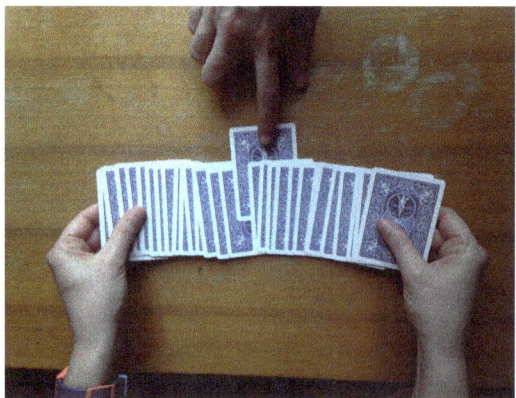

Fig. 2

6. As the spectator removes his card, separate the deck at the point from which the card was taken; see Figure 3. Tell the spectator to look at his card and remember its identity. As he does this, place the packet of cards in your right hand on the table.

Fig. 3 Photos: Brian Rappert (29 March 2018).

7. Tell the spectator to replace his card on top of the packet that lies on the table. Your instructions should be something like this: "Please place your card back in its original position in the deck." As you speak, point with the right hand to the tabled packet. As a matter of fact, the

> spectator is *not* returning his card to its original location, but this fact is never questioned.
>
> 8. When the spectator has placed his card on top of the tabled packet, place the packet in your left hand on top his card. Tell the spectator to carefully square up the deck. His card is apparently lost in the deck, but really it lies directly below the Key Card, the 3D in our example.
>
> 9. Now begin to deal cards off the top of the deck, turning them face-up as you deal. Explain that if the spectator names the cards as they are dealt, you can determine which card is his no matter how he tries to disguise his voice. Encourage him to announce each card in a different manner; he can speak in a dialect or an obscure foreign tongue; he can shout, scream or whisper. The more variety he uses, the more impossible the trick seems.
>
> 10. All you need to do is wait until the 3D shows up. Then deal the next card. This will be the spectator's chosen card, and you announce it as such.
>
> * The Three of Diamonds. This standard form of reference, with numeral and initial suit name, will be used in the book from time to time.

Consider, then, one way of making sense of this entry.

A noticeable feature is its two-part organization: Paragraph 1 of *No-Clue Discovery* sets out a performance from a third-person perspective. More than just being a fly on the wall observing what is taking place, readers as aspirant performers are invited into witnessing shared nescience: the pack is such that 'no one—not even the magician— knows where the card is located. The identification of the chosen card in the third paragraph (without any details suggesting how this could be done in the second or third paragraphs) sets the basis for a mystery. Despite being lost to everyone, the magician finds the card nonetheless.

But more than just presenting an effect unfathomable to the audience, the wording in the second and third paragraphs presents a puzzle to the aspirant reader. The amateur magician is to somehow identify the card based on the tones of utterances of the spectator—even as the instructions in the third paragraph suggest that the details of what is spoken do not matter.

With seemingly no apparent way to make sense of how the card identification was accomplished up until this stage, paragraphs four to ten then give the 'how to…' methods, speaking directly to readers. They specify that the methods at play are unrelated to the calling out of cards mentioned in the directions. Instead, a known card marks the position of the chosen card.

A feature of *No-Clue Discovery*, then, is that rather than setting out a single perspective for understanding performance, the wording provides varied ways of relating to what takes place. In this, *No-Clue Discovery* is arguably in line with many other written and face-to-face forms of direction. As Graham Jones argued in *Trade of Tricks*, imagining what spectators are seeing and thinking is a central skill honed during face-to-face conjuring tuition. Hand gestures, verbal patter, bodily movements and other actions need to take spectators toward preferred understandings and discourage others.[1] Ensuring this outcome requires performers to be able to adopt the perspectives of others.

It is hardly surprising, then, that conjurors have developed varied forms of writing designed to attend to spectators' perceptions. Scripting performances, as happens with TV dramas (for instance) is one way to foreground what conjurors wish their audiences to perceive and to remember.[2] Even the basic vocabulary favored by magicians for describing conjuring speaks to the importance of how the audience perceives what is taking place. While *Self-Working Card Tricks* adopts the commonplace term 'tricks' to label the feats set out, 'effects' are often portrayed as the prime preoccupation for magicians. *Effects* refers to what the audience perceives through the overall presentation. *Method* refers to the means and techniques whereby the effect is produced.[3]

---

1 Jones, Graham. 2011. *Trade of the Tricks*. London: University of California Press: Chapter 1.
2 McCabe, Pete. 2017. *Scripting Magic*. London: Vanishing Inc.
3 However, by no means is this distinction uniformly accepted or used consistently. For one articulation of it, see Regal, David. 2019. *Interpreting Magic*. Blue Bike Productions: 167; as well as Whaley, Barton. 1982. 'Toward a General

And, as has been argued, in magic: 'Effect should come first. Method second.'[4]

In my initial encounter with *No-Clue Discovery*, it was not just the spectator's perspective that I had to grapple with in making sense of the instructions. After reading the second and third paragraphs, I could not discern the meaning of the prior claim that 'no one—not even the magician—knows where the card' is. Was this meant as a statement of fact or a desired audience impression? In line with the expectation that magic involves extensive pretense, I was inclined toward the latter interpretation. Subsequently, I would read many trick instructions that strive to create such an impression. In this particular case, however, I came to understand the 'no-one' claim as a statement of fact (albeit one making use of a certain dramatic license in playing on the *identity* rather than the *location* of the card in the deck). In this way, I came to recognize that appreciating how to see as the performer can be a matter that needs to be wrestled with in working with directions.

However, to begin recounting my experiences with *No-Clue Discovery* as a process of reckoning with the meaning of the text in this way is already to discount the situated physical actions that accompanied my reading. As I tried out these instructions for the first time, I did not do so by reading the text from beginning to end, reflecting on ambiguous passages, settling on preliminary meanings and then picking up the cards to practice. Instead, my reaction was to physically act out the steps as I read them. When it came to the fourth and fifth paragraphs, for instance, this meant recreating the actions of both the magician and the spectator: removing the cards from the case, spreading them out, picking one of them and so on. What was the case for *No-Clue Discovery* has proven to be so ever since; my reading of instructions has been invariably accomplished through some kind of concurrent physical enactment to make the text intelligible.

Let me now turn to some of the bodily and mental work associated with enacting the instructions.

Theory of Deception' *The Journal of Strategic Studies*, 5(1): 178–192, https://doi.org/10.1080/01402398208437106.

4   Kaps, Fred. 1973. *Lecture Notes*. London: Ken Brookes' Magic Place: 1.

Correspondence

To use Livingston's terminology, enacting instructions as a lived activity entailed a good deal of effort at 'correspondence'. Because words and two-dimensional figures are not physical undertakings, continual effort is needed to coordinate bodily actions with instructions. In this case, for instance, a significant amount of the corresponding entailed repeatedly visually checking the position of my hands and the cards against Figures 1–3.

Through this inspection, points of divergence became evident. As one example of a difference I noted at the time, Figure 2 (associated with the fifth paragraph) shows a small number of cards laid out with uniform distancing. Yet my first attempt at spreading an old deck on my wooden study desk resulted in a far clumpier arrangement (see Figure 4).

Fig. 4 Photo: Brian Rappert (11 December 2017).

While I noted this divergence, appreciating whether it (and others) mattered was not evident through reading the text up to paragraph six. As a result, I stopped undertaking the steps at this point to scan the instructions ahead and then re-read the description in paragraphs one to three to judge if the differences noted would affect the outcome. Once I grasped how the chosen card was located, I judged that these differences in layout would not (even if the clumpy spreading might well be regarded as, well, clumsy). I then carried on with enacting the instructions.

As a way of developing a sense of the work associated with corresponding, let me offer a contrast. One advantage with instructions that include photos, pictorial illustrations or video imaginary is that they can display a complex array of simultaneous bodily movements that would each require lengthy individual descriptions involving specialized terminology if codified into verbal or written language. As Trevor Marchand contended in a study of woodworking training, 'skilled practices and movements regularly comprise numerous actions simultaneously performed by different parts of the body, and in an immeasurable variety of possible combinations'.[5] However, language-processed instructions are:

> constrained by time-linear sequencing, making it impossible to capture the complexity of three-dimensional movement with words. Verbal instructions are necessarily impoverished because linguistic propositions can only convey information about one salient action at a time. Other simultaneous and possibly crucial actions to the movement are either eliminated from the instruction altogether or (re)arranged to follow one after the other. Propositional representations flatten three-dimensional practice into the sequential order imposed by language, thereby rendering simultaneity time-linear.[6]

In contrast, visual imagery enables multi-dimensional forms of representation that can be compressed into a single image, which would instead take many paragraphs to elaborate in a written form.

And yet, for all of the advantages of learning through visually dense instruction material, such as DVDs and online tutorials, in my experience these came with implications for the work of correspondence. I cannot recall a single case of forwarding ahead when watching a DVD or online tutorial to check on the potential relevancy of any difference between my execution and what I interpreted the instructions stipulated at a particular point. Such a fast-forwarding would be impractical. But more than this, whereas textual figures are often characterized by neatness and precision, video displays typically involve a far messier set of affairs. Within DVDs and online tutorials, cards are often not precisely aligned, finger positions move around, other physical movements

---

5   Marchand, Trevor H.J. 2010. 'Embodied Cognition and Communication', *Journal of the Royal Anthropological Institute*, 16: S112. https://doi.org/10.1111/j.1467-9655.2010.01612.x
6   Ibid.

vary in duration and distance, etc. In short, the matters of divergence between what is stipulated and what is shown are often many! Part of the competences I developed in learning through audio-visual material was to judge when such divergences can be set aside and when they call into question the adequacy of the instructions.

In these ways, what has become evident to me is that the kind of correspondence work needed for instructions varies. Part of developing skill in working with instructions is determining, among the many details presented in the instructions, whether and which kind of correspondence is required at each step. For one step, like the spreading of cards in *No-Clue Discovery*, a loose correspondence might well suffice. For the next step in this trick or for the spreading of cards in another effect, however, precise physical correspondence with instructions can be required. For example, a relatively uniform distancing between cards may be necessary for some effects so as not to show too much of their back or front faces. When this is so, hitherto taken for granted or unrecognized qualities of the cards—such as their white bordering—can emerge as vital features.

Rather than characterizing my working with the instructions as a matter of 'following', therefore, the language of 'aligning' seems more appropriate. Instead of implying adherence, it suggests making ongoing adjustments to achieve an overall line of action supporting a sought-after effect (such as card identification). Over time, as I have gained familiarity with written instructions, I have noticed myself assessing more and more which manipulations, utterances and so on are essential, and which are tangential to the desired outcome.

I was not, however, always able to 'align' loosely. For instance, *The Lazy Magician* is another entry in the book *Self-Working Card Tricks*.[7] In contrast to the two-part organization of *No-Clue Discovery*, a notable feature of this entry is the lack of any overall depiction of the sought effect. Rather than first illustrating the effect and then describing how these results can be achieved, the instructions for *The Lazy Magician* simply provide a step-by-step listing of card manipulations. These largely consist of directives that the magician needs to issue to spectators. Along similar lines, while the revelation of a Key Card method provided a basis for

---

7 Fulves, K. 1976. *Self-working Card Tricks*. New York: Dover: 8–9.

tracking the movement of the cards in *No-Clue Discovery*, *The Lazy Magician* includes no such tracking marker. It is just a series of directives. Finally, *The Lazy Magician* does not include any figures. As a result of the absence of such reference points for gauging the sought-after line of travel, the work of coordinating my actions with the instructions took on a mechanical quality. I manipulated the cards without having a sense of why or what for. I did so with the expectation that I could make sense of the reasoning for these manipulations at the end of reading the text (which only partially took place to my satisfaction). What impressed me at the time of trying out this entry was the parallel the instructions set up between the magicians' directives for spectators and Fulves' instructions to learners. In both, individuals are meant to carry out certain sequential actions—shuffling, picking, squaring, transferring, counting—but without any pointers as to why or what for. It is perhaps not surprising that while rehearsing *The Lazy Magician*, I repeatedly could not make sense of what I needed to do. Without reasons for acting, it became problematic to coordinate, correspond and undertake other work needed to put instructions into practice.

Aligning physical manipulations with instructions can become overtly question-begging in situations in which instructions include divergent prescriptions for action. In this vein, to return to *No-Clue Discovery*, have you noticed that the instructions post-replacement of the chosen card differ in an important respect? If not, have a re-read of Box 1. I only noticed it during my fourth run-through. Paragraph one asks the spectator to cut the resulting deck, whereas no such directive is given in paragraph eight. Both courses of action are possible, though the former is not without its risks. While cutting the deck further substantiates the belief that 'no one—not even the magician—knows where the card' is located, cutting risks separating the Key Card from the chosen card.[8] In the face of such recognized divergences, readers have to decide for themselves what should be done. It is just this need to consider how to go on in the face of absent, contradictory or even inaccurate details

---

8   My fear initially was that this separation might jeopardize my ability to identify the chosen card. As I realized, a *single* cut in-between the cards would result in the Key Card being located at the bottom of the deck and the chosen card at the top. This did take place once. A spectator called off all of the cards from the deck, and then the Key Card was the last one. I then knew the chosen card was the first one flipped over. By this point, though, the spectator appeared exhausted with this now lengthy display of 'magic'.

that some magicians identify as a vital *advantage* of written texts.⁹ Through their blemishes, instructions demand considered reflection and, therefore, enable future innovation.

Overall, then, as part of my development, the starting imperative to seek a close correspondence between instructions and actions gave way to conscious recognition of the scope for variation.

Envisaging

The previous subsection set out some of the work of corresponding. Enacting the instructions involved attempting specified physical actions (spreading cards, cutting a deck, making an utterance) to achieve certain positional arrangements of cards and bodies.

In the practical actions of how to make this-spread, this-cut and this-utterance, more work was taking place than just concerning the position of cards. Instead, senses of self and other were implicated.

As a way into characterizing how this was the case, consider two contrasting orientations to experience. In *Being and Nothingness*, the philosopher Jean-Paul Sartre set out this concept of 'The Look' through imagining a situation wherein:

> ...moved by jealousy, curiosity, or vice I have just glued my ear to the door and looked through a keyhole [...] [B]ehind the door, a spectacle is presented as 'to be seen', a conversation as 'to be heard'. The door, the keyhole are at once both instruments and obstacles; they are presented as 'to be handled with care'; the keyhole is given as 'to be looked through close by and a little to one side', *etc*. Hence from this moment 'I do what I have to do'. No transcending view comes to confer upon my acts the character of a *given* on which a judgement can be brought to bear. My consciousness sticks to my acts, it *is* my acts; and my acts are commanded only by the ends to be attained and by the instruments to be employed. My attitude, for example, has no 'outside'; it is a pure process of relating the instrument (the keyhole) to the end to be attained (the spectacle to be seen), a pure mode of losing myself in the world...¹⁰

---

9   Comments made during Earl, Ben. 2020, July 18. *Deep Magic Seminar*.
10 Sartre, Jean-Paul. 2003. *Being and Nothingness: An Essay on Phenomenological Ontology*. London: Routledge: 347–348.

In referring to his consciousness having 'no outside', Sartre evokes a sense of absorption in living an experience without the need to justify one's actions or even to be self-consciously aware of them.

He then goes on to contrast this scenario with what takes place when 'all of a sudden I hear footsteps in the hall. Someone is looking at me! What does this mean? It means that I am suddenly affected in my being...'.[11] As Luna Dolezal elaborates, at one level, for Sartre to be affected is to become reflectively self-aware of one's actions. As she outlines:

> once we are captured in the Look of another we suddenly separate ourselves from the activity in which we are engaged and see the activity and ourselves as though through the eyes of the other. Through this ability to 'see' oneself, afforded by being seen by another, we gain knowledge about the self, knowledge which is essentially unavailable through introspection.[12]

Yet, as she contends, this self-awareness need not require the physical presence of others. Through evoking an imagined sense of an absent or abstract Other, it is possible to see and evaluate oneself from the outside.[13]

My efforts at corresponding in the case of *No-Clue Discovery* did *not* entail the kind of selfless absorption Sartre initially described in looking through a keyhole without care for being observed. Instead, the work of correspondence was frequently undertaken with a self-awareness of my actions. I undertook bodily actions in relation to an anticipated audience, an imagined Other. This Other was scrutinizing and evaluating my efforts. While hardly unique among performing arts, anticipating what audience members see, think and feel is arguably an integral form of reasoning in magic given the importance of deception.

As I fancied at the time anyway (see the concluding section below), for me such imaginations of the Other took the form of something like a visualized video recording filmed from across my table. My card manipulations featured in the center of the frame. Through this envisaging, I anticipated others' experiences and I came to understand

---

11 *Ibid*.
12 Dolezal, Luna. 2012. 'Reconsidering the Look in Sartre's: Being and Nothingness', *Sartre Studies International*, *18*(1): 18. https://doi.org/10.3167/ssi.2012.180102.
13 Dolezal, Luna. 2012. 'Reconsidering the Look in Sartre's: Being and Nothingness', *Sartre Studies International*, *18*(1): 18–20. https://doi.org/10.3167/ssi.2012.180102.

myself through their eyes. In doing so, my own ways of perceiving were taken as the analogic model for how an absent Other would perceive my undertakings.[14] In philosophy of the mind, the term 'simulation' refers to how we attempt to know the minds of others by emulating and ascribing mental states based on our ways of making sense of the world. Using one's mind as a model for generating a sense of others' experiences can entail the conscious forming of a representational depiction, as it did so for me in the form of a running film.

Envisaging through simulation was not only at work concerning my undertaking of this or that step in *Self-Working Card Tricks*, but in relation to further anticipated audiences' responses to each step. For instance, as part of getting the chosen card under the Key Card, paragraph seven of *No-Clue Discovery* calls on the magician to verbally mislead the spectators about the return position for the chosen card. Fulves also contends that this ruse is never called into doubt. More than just the achievement of some physical action, the instructions hinge on securing an additional outcome: the non-questioning of the card placement by the spectator. In this respect, as with many other trick instructions, *No-Clue Discovery* provides explicit indications of how spectators will and will not respond. These form a kind of working theory of behavior. It is a theory insomuch as the instructions predict how spectators will interpret situations, posit competencies, ascribe intentionality, establish expectations and foretell reactions. It is a theory presumably distilled from Fulves' considerable experience—a know-how itself informed by the previous experiences of other conjurors.

And yet, in the case of the placement text above, my envisaging led me to doubt the wisdom of the 'Place your card back in its original position' directive. As I got to the end of paragraph seven in my first enactment of the instructions, I saw the card being placed on the packet and felt a jarring between the 'original position' verbal designation and the physical positioning. However, this was also accompanied by a recognition that what I had envisaged was based on my acquired knowledge of the methods at play. Accordingly, I tried out variations for how the cards and my hands could be positioned while imagining

---

14 See Goldman, A.I. 2002. 'Simulation Theory and Mental Concepts'. In: *Simulation and Knowledge of Action*, Dokic, J. and Proust, J. (Eds). Amsterdam: John Benjamins: 35–71. https://doi.org/10.1075/aicr.45.02gol.

how spectators would see these alterations in the absence of knowledge about the methods. Thus, at times, while the instructions provided the basis for forming my mental simulations, my simulations later also provided the terms for assessing the adequacy of the instructions.

Running through the instructions in this manner also made me appreciate how the audience's ongoing actions are *not* included in the instructions.[15] As I read through the other entries in *Self-Working Card Tricks*, my speculations about the behavior of spectators would lead to repeated concerns about the relevance and sufficiency of the details given related to the ongoing ways in which audiences would orientate, monitor and react to my doings. For instance, almost none of the tricks in the book speak to the physical positioning of the audience vis-à-vis the magician, though this issue would directly bear on matters such as the likelihood that someone could detect my attempt to see the bottom card of a deck without being noticed. Thus, I could undertake the specified steps to find the chosen card, but how my doings would be responded to at each step was uncertain.

Such realizations, in turn, would lead me to try to sharpen my awareness of what factors were at play in trying to understand the perspectives of others. I did so as part of my development by consulting the academic 'Theory of Mind' literature. Within this writing, the embodied quality of how we know each other is a recurring, though multiply conceived, theme. Philosopher Shaun Gallagher has contended that individuals might exceptionally relate to each other in face-to-face interactions by holding a theory about each other or by trying to access each other's mental states through inner simulations of reasoning. In general, though, lived interactions are often characterized by a rich diversity of ongoing signaling that provides immediately accessible evidence for others' reasoning.[16] Eye and other bodily movements, facial expressions, posture, displays of emotions and expressive actions make attempts at 'mindreading' more like 'body reading'. As such, rather than others' minds being hidden, to perceive the actions of others is to already know their meaning and intentions. Body reading in this

---

15 Whilst I performed variations of *No-Clue Discovery* on many occasions, I have never included this directive.
16 Gallagher, S. 2001. 'The Practice of Mind', *Journal of Consciousness Studies*, 8(5–7): 83–108.

sense is a capacity young children develop well before they can engage in complex hypothetical deliberations about others through explicit theories.

The result of consulting this philosophical literature for me in early 2018 was to draw my attention further to the importance of ongoing embodied signaling and the lack of regard for these matters in *No-Clue Discovery* and elsewhere.

Missing from these instructions, then, is what seems central to the undertaking of tricks: the ongoing, moment-by-moment, lived interactions between individuals. To state this is not just to contend that the instructions are no substitute for hands-on experience. It is also to point out that instructions such as that of *No-Clue Discovery* do not identify or contain all the resources needed for navigating the moment-by-moment undertaking of tricks. While, as a set of instructions, the text of *No-Clue Discovery* might provide a sense of the sought-for result of the physical manipulations, it does not provide guidance about how to make sense of the adequacy of one's action vis-à-vis the audience's expressions, positioning, and a host of other situational and emerging considerations.[17]

As I would later come to appreciate, this was not my individual concern alone. For instance, for magicians that use engagement with spectators as a basis for concealing methods, the difficulty of incorporating moment-to-moment lived interactions in instructional books can render the written medium unsuitable for teaching. [This is so, not least, because written instructions often make effects appear downright implausible unless they are also demonstrated through enacted performances.[18]

Further complicating matters, determining the adequacy of instructions depends on what counts as their 'successful' enactment. By envisaging different scenarios for how my undertaking of the placement directive in paragraph seven of *No-Clue Discovery* might be perceived, I concluded that there were a range of possibilities for what could count as success:

---

17 In missing this information, the instructions implicitly render social interactions into individual deeds.
18 Watch around 81:00 and 1:22:00 of DaOrtiz, Dani. 2017. *Penguin Dani DaOrtiz LIVE ACT*. https://www.penguinmagic.com/p/11142.

1. No one noticing that the placement was not in the original position;
2. Some people not noticing this;
3. No one explicitly pointing it out;
4. Whether or not anyone noticed or pointed it out, the audience enjoying the trick.

Yet, as a novice, I did not have any basis for assessing which of these was appropriate. Outcome 1 might seem most in line with the instructional text, and self-evidently preferable. And yet, in later years, I would come to regard all of these as potentially suitable outcomes depending on the situation at hand. In any case, at the time of first attempting this trick, the perceived absence of a standard for judging adequacy was a significant source of befuddlement. The overall sought effect hinges on the placement of the chosen card in the desired position. Indeed, the physical manipulation instructions up to that step can be interpreted as driving toward this one specific move. And yet, even while I learned to undertake the physical directions, the adequacy of my actions came into doubt because I did not have a defined sense of how to judge my undertakings.

Double Vision

While, in the past, instructional books served as an essential resource for aspiring magicians, today a vast range of audiovisual resources are available through DVDs and online platforms such as YouTube. My engagement with audiovisual instructions began in the late spring of 2018 as part of learning 'sleight of hand' manipulations through the video edition of the classic instructional book titled *The Royal Road to Card Magic*.[19] Subsequently, this self-training was complemented by watching video instructions of the sleights and tricks given in *The Royal Road to Card Magic* produced by others on YouTube. Still later, I would go on to watch instructional videos for a wide range of other sleights and routines.

---

19 Hugard, Jean and Braué, Frederick. 2015. *The Royal Road to Card Magic* (Video Edition). London: Foulsham.

As elaborated previously, in practicing written instructions, I used my imagination about what I would experience as a model for gauging others' views, feelings, apprehensions, etc. Practicing with instructional videos offers a contrasting footing. For instance, it is commonplace that tutorials start with a model performance of an effect, and then proceed to offer step-by-step instructions.[20] During the model section, learner-viewers are positioned as an audience. Whereas reading a text requires the learner to imagine what viewers will experience, videos enable learners to visually perceive and affectively react. The position of the learner changes with the viewing. With the acquired knowledge of the methods at play, learners take on the perspective of an insider who knows what to look for in scrutinizing the production of the effects.

And yet, the situation is often far more complicated than this too. Videos might relieve some need for imagination by displaying a scene, but the question of what is displayed still needs to be reckoned with. What a training video provides is not a demonstration that component sleights or culminating effects can be done in general, but a demonstration that they have been executed in a specific situation. The flipside of this specificity is that witnessing one enactment is no guarantee it can be executed elsewhere. The camera angle is the most obvious consideration bearing on whether the effect one experiences as a learner-viewer can be achieved in a different environment. Counterfactual envisioning is one way of trying to resolve what is shown.

Questions about what the video demonstrates become especially acute given the commonplace practice in online tutorials that instructors solely perform for a single camera. This point was driven home to me in practicing the trick 'Topsy-Turvy Cards'; the first entry in The Royal Road to Card Magic. Despite watching video after video, I just could not undertake the critical card overturn without prominently displaying ('flashing') a card when I was practicing in front of a mirror. Only after several days did I realize the issue was that I was closer to my mirror than all the online instructors were to their cameras. I took another step away from the mirror to change the angle of viewing and the overturn seemed undetectable.

Moreover, especially as many online tutorials are filmed by instructors themselves, they do not tend to involve other participants. As such,

---

20   In line with the opening paragraphs of *No-Clue Discovery*.

there are no additional witnesses that can validate the likely effects for other viewers. Even if online tutorials did include an audience, grounds exist for doubting the trustworthiness of others. Today, recorded magic performances are regularly subject to lingering qualms about the authenticity and genuineness of what is shown, even while conjuring itself is widely regarded as a packaged pretense.[21] Not only can visual effects be achieved through editing, audience reactions can be coached pre-performance or exaggerated through crafty video splicing.[22] The growth of video performances on social media platforms such as Instagram, TikTok and Facebook has been accompanied by repeatedly voiced concerns by professionals that cameras and confederates are responsible for more of the magic onscreen than magicians.[23]

Additional complications arise in making sense of what is shown in instructional videos through reference to what is *not* shown. For instance, multiple filming takes can be required to achieve a displayed effect, but instructions rarely acknowledge what remains off-screen. As a result, what an instructor can demonstrate through a video is not necessarily easy for anyone else (including the instructor) to duplicate consistently. Indeed, learning through watching and replicating others made me more sensitive to the many and varied potential deficiencies of real-life performances. As a result, the observable *perfection* of any single tutorial stood as grounds for doubting that I could consistently replicate what was demonstrated.

These reasons for doubts expressed in the previous paragraph are echoed and magnified within general cultural beliefs. Viewing visual

---

21   See, for instance, 'More fake reaction videos.... *Theory 11 Forum*. https://www.theory11.com/forums/threads/more-fake-reaction-videos.48889/. In the case of Zoom-based performances that have become commonplace since the outbreak of Covid-19, the lack of widespread public understanding that live Zoom video feeds can be subject to real-time manipulation has provided the basis for novel forms showing what is false. See Houstoun, W. and Thompson, S. 2021. *Video Chat Magic*. Sacramento, CA: Vanishing.

22   LeClerc, Eric. 2019. *Insider 30 September*. https://www.vanishingincmagic.com/insider-magic-podcast/.

23   And yet, in line with the duplicity which characterizes so many aspects of magic, the recording of magic performances is also held by professionals as enabling novel forms of audience scrutiny (for instance, through playback). These, in turn, demand magicians refrain from coarse means of manipulating what is seen (for instance, simply cutting out delicate moments of handling props) in favor of other, more subtle forms of obscuring which audiences are more likely to regard as enabling candid scrutiny (for instance, panning back the camera during delicate moments). See Jay, Joshua. 2020. January 9. *Presentation at The Session*. London.

imagery—such as photographs and videos—is often regarded as a fickle form of witnessing. Such imagery can be held up high as faithful and dismissed as contrived. The common expression 'seeing is believing' signals the cultural stock placed in observation. And yet, as records, visual images are also recognized as not the same as the events they seek to capture. By giving a particular line of sight or by foregrounding some objects, a photograph can mislead. Also, what is included within the image frame marks the boundaries of what has been left out—be that what has been intentionally cropped out or simply not captured at all.[24] Although purposeful manipulation of visual images has long been recognized as a possibility,[25] today, digital forms of data processing offer an array of prospects for manipulation and, thereby, generate thorny debates about the status of imagery.[26]

In response to the emergence of these kinds of considerations, for me, DVD and online instructional videos have taken on a kind of haunted quality: their efforts to display are invariably bound up with the production of what is absent.

Self-Other

> To know what our spectators are thinking during a magic effect, we must train ourselves to think like our spectators. At the highest level, this means anticipating a spectator's thoughts, words, and actions before they even occur to the spectator! — Joshua Jay, co-founder of Vanishing Inc.[27]

As subsequent chapters will explore in greater detail, the imperative issued by Joshua Jay to know spectators is a frequent refrain of seasoned conjurors. To engender feelings of awe, surprise and disbelief, magicians need to know their audiences.[28]

---

24 See, e.g., Mitchell, William J. 1994. *The Reconfigured Eye: Visual Truth in the Post-Photographic Era*. Cambridge, MA.: MIT Press; as well as Morris, Errol. 2014. *Believing Is Seeing*. New York: Penguin.
25 *Ibid.*: Chapter 4.
26 See, for instance, Kuntsman, Adi and Stein, Rebecca. 2015. *Digital Militarism*. Stanford: Stanford University Press. https://doi.org/10.4135/9781473936676 .
27 Jay, Joshua (Ed). 2013. *Magic in Mind: Essential Essays for Magicians*. Sacramento: Vanishing Inc: 104.
28 See, as well, Burger, Eugene [n.d.]. *Audience Involvement...A Lecture*. Asheville, NC: Excelsior!! Productions.

This chapter has considered some of the embodied forms of work—such as correspondence and envisioning—associated with learning from instructions. While my initial forays into entertainment magic were done alone in the sense of not being in physical proximity to anyone else, in many respects others were continuously made present. Herein, the process of trying to experience what spectators experience was integral to the basic demands of making sense of instructions.

As I have contended, the experiences of others can be positioned in multiple ways. As in the case of *No-Clue Discovery*, instructions can vary between inviting aspirant performers into a shared understanding with the audience or differentiating their perspectives from that of the audience. One of the demands of interpreting instructions is to discern what sort of relation to spectators is being called for by texts at different points. In a parallel fashion, instructions can vary in the kinds of readjustments they facilitate through the extent and nature of the information provided. An aspirant can be invited to achieve an intended effect, or can be led along a tightly prescribed course. The demands on the novice in enacting instructions can be considerable because of the need to appreciate what aspects of directions matter, as well as the need to employ standards beyond the instructions for assessing what might work. Yet it is just these kinds of appreciations that are out of reach for novices because of their lack of experience. In other words, working through *No-Clue Discovery* led to developing an awareness of what I further needed to make sense of the instructions. Thus, if the ability to find the adequacy of instructions is, to some extent, what skill is, then my encounters with *No-Clue Discovery* suggested that even the simplest magic instructions can make it clear to beginners that they possess the skills of, well, beginners.

In general terms, a magic performance is an activity undertaken between a designated performer and an audience, in which the former strives to influence the experiences of the latter. It is also predicated on the possibility that there are fundamentally different experiences existing between the two. As such, relations of unity and variance intertwine. In this chapter I recounted how I employed analogic simulation to establish how others would make sense of what is taking place. This was done, though, by also recognizing others were distinct individuals and thus able to have dissimilar affective states and perceptions. Both aspects were critical to the process of envisioning others. The relevance of both indicates the complex inter-relations of notions of self and other.

The previous two paragraphs spoke to the main concerns of this chapter; namely (i) the tensions of knowing others, (ii) experienced through enacting instructions, (iii) that entail deception.

For now, I want to close by suggesting how learning magic can entail becoming unfamiliar with one's self. During my initial working through *Self-Working Card Tricks* and other self-working books in late 2017, I was convinced that my simulations of others' experiences amounted to a rolling video with all parts in focus. Months later, when I tried to reconstruct what I had been imagining after undertaking some live performances, what was summoned up was a recollection of hazy, fragmented and darting imagery. Some things came into view—part of my shirt, the side of my hand and so on—but there was nothing like a 'picture frame' image in my mind. Even if I try simply to imagine what I look like from across my desk while typing these words, if I closely attend to what is summoned I notice that I cannot generate anything like a typical perceptual experience of watching the television. Try it yourself.

In this way, in being prompted to reflect on my initial 'simulations', I could not 'see' what I thought I had imagined while practicing with *Self-Working Card Tricks*. Not only did I come to question whether my experiences could serve as an analogic model for others, I also came to question whether my experiences were how I had previously understood them. Like a well-executed illusion, the blurring of perception and imagination was both befuddling and exhilarating.

3. Control and Cooperation

> The intended dupe of the magician's wiles is, of course, the spectator. He is the objective. All of the performer's endeavor is aimed at deceiving him [...] In him are combined the formidable barriers the deceiver must breach and the very weaknesses that make him vulnerable. It is the magician's task to learn how to avoid the barriers and to attack the weak spots.[1]
>
> —Dariel Fitzkee, *Magic by Misdirection*

Not least because of the deception at play, magicians frequently reflect on how they do and should relate to their audiences. Within such discussions, 'control' often figures as a prominent theme. As in the above 1945 quotation from Dariel Fitzkee, conjuring can be portrayed as an asymmetrical activity in which the audience's imagination is—or certainly should be, in the case of a competent performer—sculpted by the magician's hands. This is so because the ultimate ability in magic is *'to influence the mind of the spectator, even in the face of that spectator's definite knowledge that the magician is absolutely unable to do what that spectator ultimately must admit he* [sic] *does do'*.[2] This influencing must be secured whatever the composition of the audiences—their gender, ethnicity, occupation or any other characteristics. While individual audience members might give a magic effect their own meaning, the magician strives to convince everyone that something inexplicable has taken place.

Some of those academically theorizing about conjuring have, likewise, treated it as a contest of strength and wit. As the social psychologist Nardi contended:

> The process of performing a magic trick involves a kind of deceit that involves power, control, and one-up-man(*sic*)ship. Magic is an

---

1 See Fitzkee, Dariel. 1945. *Magic by Misdirection* Provo, UT: Magic Book Productions.
2 *Ibid*. (emphasis in the original).

aggressive, competitive form involving challenges and winning at the expense of others [...] It is creating an illusion that involves putting something over someone, to establish who is in control, and to make the other (the audience) appear fooled.[3]

Herein, while it might be readily recognized that conjuring involves actions by both the magician and the audience, the agency should rest squarely with the former.[4] The latter's role is limited to one of possessing background knowledge, perceptual limitations and social expectations that can be led this way and that.

Absent the antagonistic overtones, Simon Aronson has spoken to the imperative to mold spectators' senses: 'A magician's paramount goal is to manipulate the spectator's mind and senses to bring about [a] state of impossibility'.[5] The philosopher Leddington likewise characterized the magician as one that '*coerces* the audience into trying to imagine how the illusion of the depicted event might be produced and the main point of the performance is to *prevent* them from succeeding'.[6] One way that coercion is achieved is by convincing the audience a definite series of actions has been undertaken, but then confounding their expectations for what outcomes result.[7] Another way coercion can take place is for the audience to feel free in their choices, even when these actions are immaterial to a planned outcome.[8] Both forms of coercion rest on creating a split between the presented story of what is taking place and the real situation that makes use of hidden methods.[9]

---

3   Nardi, Peter M. 1988. 'The Social World of Magicians' *Sex Roles*, 19(11/12): 766. 'Sic' in original.
4   One entertaining guide for how to ensure this is so, see Hopkins, Charles. 1978. *Outs, Precautions and Challenges for Ambitious Card Workers*. Calgary: Micky Hades.
5   Aronson, Simon. 1990. *The Illusion of Impossibility*: 172.
6   Leddington, Jason. 2016. 'The Experience of Magic', *The Journal of Aesthetics and Art Criticism*, 74(3): 260—italics in original. For another academic analysis of the importance of control, see Jones, Graham and Shweder, Lauren. 2003. 'The Performance of Illusion and Illusionary Performatives: Learning the Language of Theatrical Magic', *Journal of Linguistic Anthropology*, 13(1):51–70.
7   For a classic statement on these themes, see Neil, C. L. 1903. *The Modern Conjurer and Drawing-Room Entertainer*. London: C. Arthur Pearson.
8   See Pailhès, A. and Kuhn, G. 2020. 'Influencing Choices with Conversational Primes', *Proc. Natl. Acad. Sci.*, 117: 17675–17679. https://doi.org/10.1073/pnas.2000682117; and Pailhès, A. and Kuhn, G. 2020. 'The Apparent Action Causation', *Q. J. Exp. Psychol.*, 73: 1784–1795. https://doi.org/10.1177/1747021820932916.
9   For a discussion of given and hidden stories, see Smith, W. 2021. 'Deceptive Strategies in the Miniature Illusions of Close-Up Magic' In: *Illusion in Cultural*

Although others have acknowledged the need for control, it has not been regarded as an unqualified objective. Thun Helge spoke to this point in an article for *Genii* magazine, by stating:

> As magicians we also like to control everything: cards, the audience... control, control, CONTROL! [...] Magic is always about control. But then again: being an artist is not about control — it's about freedom. Freedom of constraints, of obligations, expectations and worries. How is an audience supposed to feel free and liberated when the performer himself is a control freak with obsessive-compulsive disorder?[10]

In a similar qualifying vein, in *Strong Magic*, Darwin Ortiz offered 36 laws to fellow magicians. The final of these was: 'Always remain in control'.[11] A prime area identified for control was audience challenge. Ortiz advised fellow magicians to ignore hecklers proffering explanations for conjurors' feats; refuting their claims would only serve to encourage further disruption, since hecklers crave attention.[12] And yet, against the voiced imperative to cut off any contest, Ortiz also noted that some challengers were not motivated by the desire to make trouble. When audience members express reasonable suspicion about the methods for an effect, letting this air can be productive. It provides an opportunity for receiving feedback on what needs altering, and for engaging audiences in ad-lib conversation.[13] And further still, Ortiz also cautioned that allowing any interruption would likely encourage unwanted ones.

The potentially tangled relation between control and challenge has been spoken to elsewhere. Pit Hartling advocates harnessing audience challenge by encouraging it at strategically planned moments. Through 'induced challenges', what appears to be a genuine contest by spectators can, in reality, function as a means of exercising control. Conspicuously place a torn-up card on a table, for instance, and audiences may demand it be restored magically.[14] 'Voilà! Here is your restored card', says the prepared performer.

     Practice, K. Rein (Ed.). Routledge: 123–138.
10   Thun, Helge. 2019. 'Control', *Genii*. December: 71.
11   Ortiz, Darwin. 1994. *Strong Magic*. Washington, DC: Kaufman & Co.: 437.
12   Ibid.: 420–422.
13   Ibid.: 425–426.
14   Hartling, Pit. [2003] 2013. 'Inducing Challenges'. In: *Magic in Mind: Essential Essays for Magicians*, Joshua Jay (Ed.). Sacramento: Vanishing Inc.: 105–112.

This chapter describes how I came to understand the place of control in conjuring. In line with the overall approach in *Performing Deception*, I do so by considering how control is bound up with, and dependent on, one of its notional opposites. Specifically, in this chapter, I examine the interplay of and complementarity between relations of control and cooperation. In my encounters with others, the emphasis placed on 'control' in some of the characterizations of magic above seemed lopsided, investing too much agency with the magician. Investigating how relations of both control and cooperation clash and co-exist in small group interactions will serve as another way for approaching entertainment magic as *deft contrariwise performance*.

Methods for Appreciation

In preparation for moving from practicing alone to performing for others, in early 2018 I began reviewing the academic literature on entertainment magic. I hoped to locate observations and reflections on such topics as: the expectations of audiences, their inter-personal group dynamics, as well as how audiences interpret performances of magic. Relating to the social sciences and humanities, at least, what struck me at the time was the relative dearth of such literature.[15] Audiences' first-person experiences and reasoning were instead largely taken as known from their overt behaviors, stipulated by seasoned magicians, whose virtuosity was taken to imply that they can account for spectators' lived experiences,[16] or reconstructed from limited historical records. The result was a curious situation: the audience was both typically deemed central and rendered marginal.[17] Charles Rolfe spoke to this point in these terms: 'We know that magic requires a spectator, but we do not know what a spectator is'.[18]

---

15   For excellent analyses of magic that seek ways of bringing the audience in, see Jones, Graham. 2011. *Trade of the Tricks*. London: University of California Press; and Jones, Graham. 2012. 'Magic with a Message', *Cultural Anthropology*, 27(2):193–214. https://doi.org/10.1111/j.1548-1360.2012.01140.x.
16   For instance, their first-person reasoning, affective states, expectations, motivations, etc.
17   For a sustained effort to engage with audiences regarding more supernatural forms of magic, see Hill, Annette. 2010. *Paranormal Media: Audiences, Spirits, and Magic in Popular Culture*. London: Routledge.
18   Rolfe, Charles. 2014. 'A Conceptual Outline of Contemporary Magic Practice'. *Environment and Planning A*, 46: 1615.

In early 2018, I started to video small group performances, with the primary intention of understanding participants' lived experiences. I began where those new to magic often begin: doing routines for small groups of friends and acquaintances. Most of these took place around a kitchen table in what amounted to something of a blend between research and entertainment.

In the end, I recorded 30 sessions over sixteen months. Four different themed variations were put on. Each session lasted between seventy minutes and two hours. The 69 different participants were largely university faculty, academic visitors or PhD students who, at the time, were associated with universities in the UK or Sweden.[19]

In order to explore participants' experiences, the sessions departed from standard performances. Akin to a focus group, they combined the presentation of information (in this case, the effects) with moderated discussion.[20] I modified the questions and overall composition of the sessions on an ongoing basis in order to make my emerging reflections on performing magic into topics of conversation within the sessions (see Chapter 7). The expectation with this format was that, as in focus groups more generally, it would provide an open but directed space for participants to generate their questions and concerns. Furthermore, the emergent dialogue between participants would lead to novel insights, compared with interviewing individuals afterwards or asking them to fill in an evaluation form.[21]

In this chapter, I am going to pay particular attention to the first 13 sessions. Not only were they formative in my development, as a complete novice to the world of magic I could lay no claim to possessing refined skills or abilities at the time. My status as a novice will be relevant in the analysis that follows. All of the tricks performed in this first set were of the self-working variety covered in Chapter 2, whereas the remainder of the 30 sessions included self-working and sleight of hand-based

---

19 For further details of the research design, see Rappert, B. 2021. '"Pick a Card, Any Card": Learning to Deceive and Conceal—With Care', *Secrecy and Society*, 2(2). https://doi.org/10.1177/1468794120965367 and https://brianrappert.net/magic/performances.

20 For further information of the composition and rationale for these sessions, see the 'Going On' entries at https://brianrappert.net/magic/performances.

21 See J. Kitzinger and Barbour, R. 1999. 'Introduction'. In: *Developing Focus Group Research*, R. Barbour and J. Kitzinger (Eds). London: Sage; as well as Morgan, D. 1998. *Focus Groups as Qualitative Research*. London: Sage.

tricks. In terms of style, the first 13 sessions were framed through the notion of 'embodiment'—participants were asked to look in particular directions and say certain kinds of things (for instance, call off cards). In my accompanying verbal patter and bodily movements, I suggested that I was identifying selected cards based upon reading facial expressions, postures, eye movements, voice and the like. I was not.

Control and Cooperation

This section explores some of the ways both control and cooperation figured within our interactions. What follows is largely a description of *what* took place. Chapter 7 turns squarely to addressing how magic *should* be done through juxtaposing the notion of control with care.

Certainly, it is possible to identify ways in which the notion of cooperation seems of limited relevance to our interactions. Grice, for instance, has suggested that cooperation is underpinned by the belief that others are generally telling the truth, or at least what they believe to be true.[22] In the manner in which entertainment magic is regarded as entailing forms of deception, however, this starting presumption was repeatedly subject to explicit doubt. As one participant stated, 'the thing about the magic is... that the magic is not what it seems. So if the magician starts telling you they are reading a book about body language, I immediately think it's not about body language' (Session 4, Participant 1). This expressed contrarianism points toward the multi-layered and complex processes of deception-discernment at work. As a magician, I sought to anticipate the responses of participants, to factor them into the staging of the effects (for instance, to prevent detection of the underlying methods), and to riposte backchat (for instance, to reply to expressions of suspicion about my explanatory patter). Participants anticipated acts of misdirection in general and, at times, sought to see through the actions conducted. This was done, in part, based on the very details of gesture, voice, movement and so on that were meant to mislead them.

Likewise, too, it is possible to identify ways in which the notion of control was of central relevance to our interactions. As with many types

---

22 Grice, Paul. 1989. *Studies in the Way of Words*. Cambridge, MA: Harvard University Press.

of conjuring, all the sessions relied on direct audience participation in response to my directives: selecting cards, shuffling the deck, calling off numbers, etc. In this, my sessions shared in the decidedly asymmetrical relations characteristic of magic: magicians routinely state directive after directive to participants, whereas participants do not do the reverse. Magicians also conventionally exercise asymmetrical rights to speak. For instance, pauses in their verbal patter typically are not taken by onlookers as possible conversation entry points, but are instead orientated to as temporary stoppages. This was generally the case in my sessions as well. Moreover, unlike as is commonplace for other social activities (childcare, to name one example),[23] I was not compelled to escalate directives into imperative demands because individuals refused to comply with my instructions.

As another dimension of control, at least initially within each session, participants routinely described themselves as mere spectators. After some initial tricks, my questioning across all the sessions included asking participants how they thought the magic was being accomplished. Responses squarely focused on *my* actions (for instance, the belief that I was covertly manipulating cards, directing attention, etc.), with almost no regard for their role in the unfolding interaction.

In the ways identified in the previous paragraphs, magicians frequently assume an authority that would be out of place in many other settings. And yet, despite how control can be positioned as germane and cooperation as not, in the next sub-section I will advance a more nuanced understanding of how the two interplay together.

The Chemistry of Control and Cooperation

As an initial observation, participants generally did follow my directives. However, this is not all they did. On some occasions they undertook actions such as secretly removing cards, demanding to inspect the deck before and after card revelations, taking the cards away from me mid-trick so to rearrange them, or grabbing away my written notes. In an exceptional (and memorable) session, one participant undertook all of these interventions. Such interventions were disruptive in that they significantly undermined the prospect that the cards could be identified,

---

23   *Ibid.*

or threatened to reveal the underlying methods. Whilst hardly welcomed by me at the time, such exceptional interventions were crucial in raising my awareness of the extensive range of behavior forgone by most other participants.

More common than these interventions were non-compliant responses or requests that did not fundamentally undermine what could be defined as the overall 'directive trajectory' (and presumably were not intended to do so).[24] Momentarily feigning an alternative card selection, asking me to physically re-position myself, politely requesting whether they could inspect the deck, alternating the pitch of their voice, etc. were some (often playfully delivered) acts of non-compliance.

When questioned about their (typically) restrained challenges, in eight of the first thirteen sessions, participants overtly accounted for their (in)actions through appealing to their desire to contribute toward the success of the effects. One discussion unfolded as in the excerpt below. The excerpt introduces a number of transcription conventions that will be used in this book to convey nuances of talk. See pages 10–11 for further details. At this point, however, let me note some of the conventions: double parentheses denote my own summaries of what took place; single parentheses denote my best guess at what was said or the duration of pauses; the equals sign indicates words spoken without an intervening silence; underlining signals emphasis; and capitalization indicates increased volume.

Excerpt 3.1—Session 6

No	Direct transcript
1	P1: Of course I know I could mess up your=
2	P3: Yeah
3	P2: =trick.
4	P3: Yeah
5	P1: But that's not fun.

---

24 Goodwin, Marjorie Harness and Cekaite, Asta. 2012. 'Calibration in Directive/Response Sequences in Family Interaction', *Journal of Pragmatics*, 46(1): 122–138. http://dx.doi.org/10.1016/j.pragma.2012.07.008.

No	Direct transcript
6	P3: I know, I am like that as well, you know, I just, in fact I still don't want to know how he makes it because=
7	P1: Yeah
8	P3: =it's fun. I agree, you know, it is a cooperative enterprise so what's the point of
9	((side discussion))
10	P2: But I also don't think you don't want to be too disruptive because you want (2.5) you want him to succeed as well. Do you know what I mean?
11	P1: Yeah
12	P2: Like you kind of, when he spins over the card you want it to be the right card=
13	P1: =So in that
14	P3: Yeah
15	P1: =sense we
16	P2: Yeah
17	P1: =are a willing audience, but I think generally audiences (1.2) for magic at least are willing.
18	P2: Yeah
19	P1: Cooperative
20	P3: Yeah, yeah. Yeah it is a kind of a (.) game you play together. In a sense=
21	P1: Yeah
22	P3: =you don't want to be (.) disruptive. In a=
23	P2: Hmm
24	P3: =way you want to be surprised. You know=
25	P2: Yes
26	P3: =you WANT (the trick to come). You WANT to be amazed, that's the deal.

In expressing the desire for the trick to succeed, the participants spoke to the enactment of a situation in line with magician Darwin Ortiz's expectation that nearly:

> any audience may fall into the mindset of viewing a magic performance as a win-lose situation if you encourage them to. It's your job to make them see it as a win-win situation [...] A good magic performance is a cooperative venture, not a competitive one. The audience should actually be your allies in fooling them.[25]

For Ortiz and many other magicians, a vital requisite social skill is the ability to induce cooperation in others. Like managers or political leaders, conjurors do this through their movements, comportment and the stories they tell.[26] Taken as representations of motivational states, the exchange in Excerpt 3.1 serves as evidence for this session achieving a win-win situation. Yet, importantly, this cooperation was not solely achieved because of my agency. In other words, it was not my job alone. Participants retained a sense of control through the options they elected *not* to pursue in this situation and accounted for their inaction with the label 'cooperation'.[27]

Further along the lines of treating the audience as active in their own right, participants engaged in numerous forms of behavior that worked toward the accomplishment of the effects. For instance:

- They routinely used visual scrutiny, verbal corrections and pointing gestures with one another to ensure actions were taken per my directives. This was particularly important when I turned my back or left the room.[28]
- Participants monitored each other regarding the appropriateness of behavior. They verbally (and often playfully) sanctioned each other (or themselves) when a

---

25 Ortiz, Darwin. 1994. *Strong Magic*. Washington, DC: Kaufman & Co.: 22.
26 Fligstein, Neil. 2001. 'Social Skill and the Theory of Fields', *Sociological Theory*, 19(2): 105–125. https://doi.org/10.1111/0735-2751.00132.
27 The manner in which failure can be uncomfortable for an audience is discussed in Landman, T. 2018. 'Academic Magic: Performance and the Communication of Fundamental Ideas', *Journal of Performance Magic*, 5(1). https://doi.org/10.5920/jpm.2018.02.
28 Also frequent were participants' queries to me checking whether they were undertaking appropriate card manipulations.

line of action was deemed to have been taken too far or not far enough (for instance, when a participant was judged as not paying sufficient attention).

- They verbally described their actions when manipulating cards so that others would be able to follow along with the sequence of what was taking place.

- When efforts were interpreted as having 'gone wrong' (for instance, I was not able to identify the chosen card), participants offered apologies about their own shortcomings in executing the instructions.

In short, through such words, gestures, movements and postures, participants coordinated their actions with the actions of others present. More than the equivalent of the type of responsive coordination that takes place in figuring out where to stand in an elevator with strangers, they coordinated their actions in ways that entailed actively working *together* in sustaining a *shared* enterprise.[29] Furthermore, individuals engaged in varied forms of corrective behavior—sanctioning, rebuking, justifying, reminding, pointing, apologizing and so on—that worked toward sustaining their sense of what ought to be taking place.

The display and direction of attention provided another area for cooperation and the exercise of agency by participants. Attention is a topic at the fore in theorizing magic. Indeed, its manipulation through talk and non-verbal action (such as the direction of the magician's gaze)[30] is often portrayed as a central task for conjurors. As a beginner, though, what was unmistakably evident from these sessions was that participants acted together in ways that were not the result of some intentionality on my part. As in other types of small group interactions, in these sessions, ordinary forms of mutual engagement between participants (and thereby away from me) were general features of interactions. Participants watched each other, looked back at others watching them, physically orientated toward one another (for instance, during laughter),

---

29 In such respects, magic differed from the kinds of co-present coordination elaborated elsewhere, as in Goodwin, C. 1995. 'Seeing in Depth', *Social Studies of Science*, 25: 237–274.

30 Kuhn, Gustav, Tatler, Benjamin W. and Cole, Geoff G. 2009. 'Look Where I Look', *Visual Cognition*, 17(6/7): 925–944. https://doi.org/10.1080/13506280902826775.

and so on. Such actions promoted mutual regard between individuals, but undermined the prospects for all present to have a single joint focus for attention. In other words, unlike in some social activities,[31] directing gaze elsewhere than toward the notional focal activity (such as my body or the cards) was not necessarily treated as an accountable deviation from expected forms of behavior.[32] Indeed, establishing a shared visual focus by participants to the card manipulations[33] was a demand on me from time to time, especially when I wanted participants to attend to specific actions in order to foster certain memories. Conversely, at other moments participants used the words, gestures and gazes of interaction to momentarily produce shared foci for attention.

Relatedly, a common assumption in the study of magic is that audiences want to know how effects are achieved and act to decipher the underlying methods.[34] Yet, when asked whether they wanted to know the methods at play, a diversity of responses were offered.[35] Whether and what participants wanted to know were reported as turning on whether the affective value of trickery would be enhanced by knowing, whether they might be more at ease with the comfort of ignorance, and whether I could be trusted to provide a true explanation after all of the subterfuge on display. One person characterized the complexity of his orientation to knowing and being fooled in this manner:

> I think it is tricky because umm, you don't want to be, umm, fooled, I mean you don't wanta (.) miss something obvious. But at the same time, you like it when it is pulled off. So, OK, so you wanta be kind of lured by the trick but you of course don't want to be sheepishly foolish. But, of course, you won't kinda want to be all, don't you trick me, because it is

---

31 Rouncefield, M. and Tolmie, P. 2013. *Ethnomethodology at Play*. London: Routledge.
32 In an extreme instance of disengagement, one participant repeatedly attended to his mobile phone, a practice eventually verbally sanctioned by another participant.
33 As in the perceptual intersubjectivity noted by Zlatev, Jordan, Brinck, Ingar and Andrén, Mats. 2008. 'Stages in the Development of Perceptual Intersubjectivity', *Enacting Intersubjectivity*. Amsterdam: IOS Press: 117–132.
34 For instance, Danek, Amory H., Fraps, Thomas, von Müller, Albrecht, Grothe, Benedikt and Öllinger, Michael. 2013. 'Working Wonders? Investigating Insight with Magic Tricks', *Cognition*, *130*(2): 176. http://dx.doi.org/10.1016/j.cognition.2013.11.003 176.
35 In line with Jay, J. 2016. 'What do Audiences Really Think?' *MAGIC* (September): 46–55. https://www.magicconvention.com/wp-content/uploads/2017/08/Survey.pdf

part of the sensation that you are going to be tricked. So I think it is kinda of double. You both want and don't want to be fooled (S12, P2).[36]

Moreover, participants also reported more deliberate kinds of modulated attention. For instance, intentionality was brought into play through deliberate efforts to *disengage*:

Excerpt 3.2—Session 13

No	Direct transcript
1	P4: I guess in my case I tried to not look at the card, too much, ahh when you were doing the trick with me, umm, I won't not look at it, but look at all the cards, equally, kinda shifting a looking at you a lot, where you're looking. But when, umm, in the other cases I just tried not to get involved, because I did not want to give it away. Like I did not listen to ((P2)) when she was counting, I did not know her card. umm
2	BR: OK
3	P1: So you were afraid that you would give=
4	P4: Yeah
5	P1: =the answer away when
6	P4: Yeah
7	P1: Ah, OK.
8	P4: If I knew her card then maybe I was going to look at it too much and he (.) would see that.
9	P1: AH.
10	((Group laughter))

These comments point to how participants attempted to exert agency within situations by intentionally directing attention *away* from the performance.

---

36 For a further discussion on how audience can be 'torn between the enjoyment of belief and the resentment of being fooled' see Neale, Robert E. 2009. 'Early Conjuring Performances', In: E. Burger and R. E. Neale (Eds). *Magic and Meaning* (Second Edition). Seattle: Hermetic Press: 43.

Overall, instead of a one-way process of control by the conjuror, the considerations noted in this section suggest a more negotiated, multi-directional dynamic. While participants undertook various forms of non-compliance that could be regarded as opposing my efforts at control, these were intermixed with actions that helped maintain the setting as one of the performance of magic, and furthermore were frequently orientated to by participants as instances of intentional cooperation.

Accounting for Control and Cooperation

The previous section examined the interplay between control and cooperation, in part through reproducing participants' statements. As with most forms of social research, in the case of these conjuring sessions the methods employed were constitutive of the data produced. By my prodding through questions, participants responded in ways that went beyond the typical (dis-)affiliation displays that often follow magic effects (for instance, applause, laughter, jeers, expressions of 'How did he do that?'[37]). Instead of just being with the activity at hand, they were explicitly asked to account for their participation. The issues they voiced helped constitute a sense of the unfolding scene at hand, there and then. As conversation is a kind of collaborative conduct in the first place, the exchange of dialogue itself helped constitute a sense of the scene as cooperative.

As part of the overall dynamics, I now want to turn to how rules and norms were evoked as justifications for cooperation. Reference to rules and norms defining a sense of proper conduct for a magic performance was commonplace across the sessions. In eight of the thirteen initial sessions, for instance, participants spoke of their conscious commitment to shared standards that bounded the scope for legitimate conduct. This commitment was described at times by expressions such as that given by one participant that 'You play, of course, to the rules of the game'. Elsewhere a more elaborated relevance of norms was articulated. When asked why they had not sought to interfere with the tricks, the following discussion ensued (the placement of left square brackets on two successive lines indicates the start of overlapping talk):

---

37 For a discussion of those displays, see Ortiz, D. 2006. *Designing Miracles: Creating the Illusion of Impossibility*. A-1 MagicalMedia.

3. Control and Cooperation

Excerpt 3.3—Session 3

No	Direct transcript
1	P1: That would violate a
2	P1: [norm that, I mean, there is this sort of implicit
3	P2: [YEAH
4	P1: participatory =
5	P2: Hmm
6	P3: = expectation that we are all part of this performance and, and we just implicitly trust that, we know there is an explanation for this. There are mechanisms=
7	P2: Hmm
8	P3: =there are a logic behind this, but we want to be caught up in this and share this experience so we go along with you. We let ourselves be guided by you.
9	P2: ((side point)) We know that we are both in this
10	P3: Yeah
11	P2: together. Sort of a, so it is not like you're doing magic (.) to us.
12	P1: Hmm
13	P2: It's like we are
14	P3. Yeah
15	P2: You know, agreeing to do magic. Whether it is fantasy=
16	P1: Yeah
17	P2: =or logic
18	P2: [sort
19	P1: [Well we talked about body language too. If we were not giving you, ongoing feedback and raising our eyebrows and no way that is a good one Brian.
20	P3: ((laughter))
21	P1: If we were just a dead unreceptive participant, that would have changed the character of all of this. Certainly
22	P1: [so we play an active role in determining how

No	Direct transcript
23	P2: [So like
24	P1: this develops as well, the audience does.

Again, at one level, what is at stake in these characterizations is how individuals report on their motivations and assert agency. In this case, by being able to step back from the ongoing interactions and offer jointly formulated reflections on what was taking place, those present were able to perform a sense of themselves as knowledgeable about magic and competent to play their part as participants. In addition, rather than a state of acquiescence being secured by the magician's one-directional control, the contention was that the effects unfolded through the willingness of participants to co-produce certain patterns of relations with the magician (Lines 8, 9–15). An implication that followed was that this willingness could disappear if the participants opted for this course of action.

Although rules and norms were widely evoked across the sessions to justify behavior, the meaning of those standards was not the same between participants. When participants cited norms, they did so to render *their* behavior as that which ought to count as ordinary, expected, what anyone would do, etc. And yet, as evident in my sessions and the writing of professional magicians, audiences vary considerably in their conduct (for instance, concerning the extent they seek to disrupt magicians' verbal patter or physical actions).

In one respect, the variation in the range of activities said to be aligned with the 'norms of magic' is hardly surprising. While some games like chess have established rulebooks for gameplay—even ones for player etiquette—no such manuals exist for conjuring performances. Despite the inability of anyone to point to some established definite, written down 'rules' specifying what kinds of behavior is acceptable, participants' citations of norms gestured toward something pre-given, out-there, known, etc. In this way, appropriate standards for conduct were defined as existing separate from our interactions around a table.

Such a recourse to norms is in line with the kind of objectification Kenneth Liberman identified in how rules become orientated to as 'social facts' during the playing of board games. Rather than the produced orderliness of gameplay being regarded as the 'practical achievement of the players' concerted gameplay', he found game players accounted for rules as existing 'without any immanent connections to the players who produced' them.[38] In other words, rules were regarded as rightfully determining how a game should be played by players even as they invariably end up interpreting and negotiating the meaning of rules. Liberman's analysis itself was in line with a long history of ethnomethodological studies that note a basic disparity; the contrast between the way norms and rules are said by individuals to serve as definitive, objective standards and the way groups actively labor to establish the meaning of norms and rules.[39]

To treat norms as phenomena-in-the-making entails orientating to the invoking of norms within conversations as in itself a form of situated action. Consider some ways in which the invoking of norms can be consequential. In Excerpt 3.3, a norm was explicitly identified (Starting on Line 1). In this case, the appeal to the relevancy of norms developed a sense of the joint moral situation at hand (it was one of trust—as in Lines 6–8). The understanding promoted of the situation proved a background context for making sense of body language later (Lines 19–21).

Beyond just explicit reference to norms, when participants presented their actions and inactions as born out of commitments to cooperation, this helped influence the understanding of magic as a practice there and then. As an example, in the case of Excerpt 3.1, the reference to cooperation accounted for the behavior of participants through characterizing magic as an activity in that session (Line 10) and in general (starting at Lines 17), retrospectively offered a justification for participants' behavior. It labelled this specific interaction as a shared one of 'cooperation' by individuals that were accountable for their actions, and thereby set out a framework for interpreting what subsequently took place.[40]

---

38 Liberman, Kenneth. 2013. *More Studies in Ethnomethodology*. Albany, NY: State University of New York Press: 108.
39 See, e.g., Heritage, John. 1984. *Garfinkel and Ethnomethodology*. Cambridge: Polity Press; and Wieder, D. Lawrence. 1974. *Language and Social Reality*. Paris: Mouton.
40 A reading inspired from Wieder, D. Lawrence. 1974. 'Telling the Code'. In: *Ethnomethodology*, R. Turner (Ed.). Harmondsworth: Penguin.

64 *Performing Deception*

In such ways, our identities as audience members and as a magician were established as part of the emerging and jointly negotiated interactions. This is evident, too, in the instances that were orientated to by participants as norm deviations. Consider the following exchange. It was prompted by asking participants how they thought the effects up to that point had been accomplished.

Excerpt 3.4—Session 7

No	Direct transcript
1	P1: Well, attention is being <u>directed</u>
2	BR: By who?
3	P1: By you. Yeah, yeah.
4	BR: How have you felt me directing your attention?
5	P1: Well, because it's a contract and we are here to be entertained and in order to be entertained we know we have to play along with the rules and you are the person that is providing the rules. And so you are saying things like, umm, check these cards, now have a good look at them.
6	BR: Uhm
7	P1: And it is impossible for us to do that while also paying <u>lots</u> of attention to you.
8	BR: Yeah
9	P1: So we are having our attention drawn away from where the action is going on.
10	BR: Okay, Okay.
11	BR: [Yeah
12	P1: [That's how I have seen, and that is how whenever I have seen anything about magic that it has been explained, and that it is just amazing that you can (1.0)
13	P2: Draw attention
14	P1: Draw attention away from what you are doing.
15	((side conversation))

3. Control and Cooperation

No	Direct transcript
16	P1: Some of it, I think, is physical manipulation, and you having a chance to look at cards and re-arrange cards in interesting ways. But in order to do that surreptitiously our attention has to be elsewhere. And you have to have (.) quite a lot of physical dexterity. And it's like playing a musical instrument and singing, or, you have got to do more than one thing at a time. So you have to get the patter going=
17	BR: Yes, yes, yes
18	P1: =and sound really confident as well as the fact that you are surreptitiously looking at what the bottom card is because in a lot of these tricks, sorry am I saying too much?
19	BR: No, no, no, it's fine. Whatever.
20	P1: But a lot of tricks, what's happening is that the card is either being placed on the top or the bottom, but seems to be concealed, but is in a <u>prime</u> place and that means as long as you have enough dexterity, you can ((inaudible)) make sure you know roughly where it is.
21	P2: But he needs to know more than roughly. Don't you?
22	P1: But we need to then not be distract, we need to be distracted. In some of the tricks it is easier to see that happening than others.

Herein, the participants and I unfolded a sense of the scene together through our verbal exchanges. What P1 perceived in our encounter was spoken to through reference to her prior familiarization with card tricks. In Line 5, she suggested that performances entailed a contract between audiences and magicians in which the former play to the rules of the latter. Across Lines 12–14, she compared her experience in this session to previous encounters with magic, and grounded her statements based on past exposure to explanations of secreted methods. Such utterances functioned in a real-time manner to develop a sense of the scene at hand and the identities of those in it. They framed our interactions in terms of distinct roles; authorized me to act as the performer; gave a gloss of our previous interactions, as in line with conventional roles; accounted for limitations in being able to specify the detailed methods of the tricks; and provided a resource for making sense of later interactions.[41]

Subsequently in this exchange (Line 18), P1 would offer a general-level description for the methods for effects—an act that she orientated to as transgressing the proper audience role spoken to in Line 5. In subsequently seeking approval for this action (Line 18), she sought to repair any perceived transgressions. In this way, both a sense of the specific scene as well as the nature of magic as an activity was worked up through the exchange.[42]

Dialectics of Control and Cooperation

This chapter has analyzed my initial experiences in performing magic in 2018–2019 through considering the place of control and cooperation within them.

While certainly not denying entertainment magic often entails efforts to control the thoughts and behavior of audiences, the analysis presented in this chapter has given reasons for questioning: (i) binary oppositions between the magician and their audience, and (ii) tendencies to reduce performances to the doings of the conjuror. As a result, during 2018 I came to understand magic as a form of what can be called 'reciprocal action'. Reciprocal action refers to situations in which 'changes in one [person leads to] changes in the other, and the process goes back and forth in such a way that we cannot explain the state trajectory of one without looking at the state trajectory of the other'.[43]

When approached as a reciprocal action, space opens up to move away from conceiving of conjuring solely as a one-directional exercise

---

41 For an analysis of how norms and situations are mutually constituted, see Wieder, D. Lawrence. 1974. *Language and Social Reality*. Paris: Mouton.

42 Following Goodwin, we can treat these kinds of sense-making efforts as emergent 'co-operative' undertakings. Co-operative here designates how individuals produce actions on the basis of reusing and transforming the discursive resources provided by others. In this exchange, we were cooperating with each other through varyingly relating to each other's utterances—for instance, by explicitly drawing on one another's statements (Lines 17–18), by expressing doubt about others' contentions (Lines 34–35), and generally by offering statements designed in response to others' prior conversation. See Goodwin, Charles. 2017. *Co-Operative Action*. Cambridge: Cambridge University Press.

43 Kirsh, David. 2006. 'Distributed Cognition', *Pragmatics & Cognition*, 14(2): 250. https://doi.org/10.1075/pc.14.2.06kir.

of control by an individual secret keeper. Instead, it becomes possible to orientate to it as a moment-by-moment negotiated ordering between all of those present, organized together by all those present. Herein the actions of an individual audience member need to be understood through their situated and embodied relation to the magician and other audience members, and the magician is understood through their situated and embodied relation to members of the audience. In the case of my sessions, reciprocity was relevant both within the group dialogues, as well as within the performance of the effects.

Treating magic as entailing reciprocal action, though, does not in itself resolve how control and cooperation *will* or *should* interplay together in any specific encounter. As noted previously in this chapter, control of audiences' thoughts and behavior is frequently portrayed by magicians at times as an unqualified imperative. As such, it should be maximized. Control is what enables feelings of astonishment, excitement and wonder. Alternatively, at times, control has been positioned as needing to be balanced against other considerations. For instance, reining in magicians' will to control can encourage spontaneity and connection.

In seeking to describe the interactions in these sessions, my goal has not been to advance an argument as to what counts as the proper mix of control and cooperation for conjuring. Instead, I have sought to draw on the details of the interactions to make a more preliminary argument: how control and cooperation can mutually depend on and contribute to each other as part of phenomena-in-the-making. In particular, as a response to the emphasis often given to control by seasoned magicians, I have attended to my experiences and reflections as a beginner without any extraordinary ability to influence others. It is from this status as a novice that I developed an awareness of how audiences engaged in forms of cooperation that worked towards the mundane but vital practical tasks: ensuring directives are followed; shifting attention away from myself; producing joint objects for attention; looking at effects but not watching for methods, and so on. While this section has sought to elaborate how the methods employed for promoting dialogue were constitutive of the data produced, Chapter 7 will go on to consider how I would later come to marshal this condition within the design and delivery of shows for the public.

Coda

I started this chapter with an observation of how the experiences of audiences both feature as central to, and can be marginalized in, attempts to understand conjuring. Through integrating reflection on the interactional dynamics of magic within performances, the sessions considered in this chapter were intended to make individuals' implicit feelings and experiences into explicit topics for group conversation. In seeking to provide an analysis attentive to details of our interactions, I aimed to take (co-)participants' accounts of experiences seriously.

However, doing so has relied on an underlying premise: namely that participants' accounts can be taken at face value. In other words, this analysis has assumed that others were ordinarily telling the truth, or at least what they believed to be true.[44] Such a starting presumption is commonplace in social life. From an ethnomethodological approach, David Francis and Stephen Hester contended that individuals:

> ... seldom have the freedom to engage in [...] idle speculation about the motives behind the actions of others. The fundamental constraint that operates in all interaction is that persons should, wherever possible, take things 'at face value'. In other words, one should respond to the actions of others on the basis of what those actions seem, obviously or most plausibly, to be. If something seems quite obviously to be a question addressed to oneself, then respond to it as such. The same holds for the meaning of what is said. If the meaning of the question is clear, then respond to it on that basis.[45]

Similarly, philosophical (and specifically phenomenological) approaches for how we know others' reasoning and intentions (see Chapter 2) are often based on the assumption that a pragmatic understanding of others can be gained by attending to the face value meaning of their overt bodily movements, facial expressions, posture, displays of emotions and other expressive actions.[46]

---

44 See Grice, Paul. 1989. *Studies in the Way of Words*. Cambridge, MA: Harvard University Press.
45 Francis, David and Hester, Stephen. 2004. *An Invitation to Ethnomethodology: Language, Society and Interaction*. London: Sage: 7.
46 See Gallagher, S. 2005. 'How the Body Shapes the Mind'. In: *Between Ourselves: Second-Person Issues in the Study of Consciousness*, Evan Thompson (Ed.). Oxford: Oxford University Press.

In contrast, my engagement with conjuring has suggested a strong dose of caution regarding what to take at face value. Magic is an activity that routinely turns on the misalignment between appearances and doings. Learning magic entails opening to the considerable potential for marshalling notions of what is obvious, plausible, on-the-face-of-it and so on through voice, gesture, eye direction, bodily movements and the like to deceive others about the state of the world. That might be about which card is in a pocket, whether this deck of cards is still the same deck of cards that was used before, etc.

More than this, audiences of conjuring generally anticipate hidden moves, lies, bluffs and other misleading acts. Yet this anticipation does not necessarily hamper the potential for magicians to mislead. Instead, it provides further grounds for it. By engaging with the beliefs and perceptions of audiences, including their suspicions about how conjurors might mislead, it is possible to exert control. The next chapter elaborates how magicians seek to marshal subtle movements, precise wording, directed gestures and many other commonplace behaviors in order that their actions appear justified to scrutinizing eyes and ears. Now I wish to attend to an alternative matter.

With all the concerted efforts toward deception on my part, it is perhaps not surprising that, over the course of putting on my initial sessions, doubt crept into my mind regarding my ability to read others and regarding the wisdom of taking their statements at face value. For instance, my attempts during the recorded research examined in this chapter to solicit critical feedback from participants generated few negative responses. As I was a complete novice who could not but improve my technical and presentational skills, the absence of criticism led me to ask: might participants be deceiving me? Could they be offering accounts of their experience that they thought I wanted to hear? Might they be speaking and acting in ways at odds with their inner thoughts and feelings?

At the time, my grounds for concern were deepened by reading two sets of literature. One, my growing familiarity with the writing of professionals gave reason to believe that at least some were wary about the ways audiences try to please magicians.[47] Reflecting on his

---

47 The Jerk. 2016. 'The Importance of Combining Methods'. http://www.thejerx.com/blog/2016/6/30/the-importance-of-combining-methods; Brown, D. 2003. *Absolute*

experience before becoming a household name in the UK, for instance, Derren Brown spoke to one dimension of audience deception:

> One problem with magic is that too often, people are polite in their responses, and we think we are getting away with methods when we simply are not. I hope you have had the experience of overhearing a spectator correctly guess exactly the method you used to achieve an effect that you have honed and worked on for years. In such situations you wonder how often this happens and you simply don't hear. But there are enough dreadful magicians around for us to know how easy it is to perform magic badly and not get any feedback. Where, after all, could that feedback come from? Not from the public, who would in most cases pretend to be fooled out of sheer pity [...] For an art that relies entirely on the experiences of the spectators, it is remarkably difficult to find out what those experiences are. We cannot finish an effect and then immediately have the audience dissect their experience of it to provide us with useful information. Yet that is exactly what we need.[48]

In its design, my sessions realized a form of the immediate dissection Brown advocated. Instead of just doing effect after effect, I engaged audiences in discussions based on what was taking place there and then. And yet, this design in itself does not bypass the basic problem of audience insincerity.

As a second literature, sociologists and psychologists have identified the ability to manipulate the truth and falsity of information as a vital skill, one learnt early in our personal development.[49] For instance, within

---

Magic (Second Edition). London: H&R Magic Books; Armstrong, Jon. 2019. *Insider* (16 December). https://www.vanishingincmagic.com/insider-magic-podcast/ and Clifford, Peter. 2020, January 12. *A Story for Performance*. Lecture notes from presentation at The Session. London.

48 Brown, D. 2003. *Absolute Magic* (Second Edition). London: H&R Magic Books. Despite his success—in television, stage and close-up forms of magic—20 years after rising to national prominence, Brown continued to argue that performers cannot judge by themselves how well shows went; as in Brown, D. 2021, May 3. *Bristol Society of Magic—Centenary Celebration: An Evening with Derren Brown*. Bristol. See, as well, Vernon, Dai. 1940. *Dai Vernon's Select Secrets*. New York, NY: Max Holden; Frisch, Ian. 2019. *Magic Is Dead: My Journey into the World's Most Secretive Society of Magicians*. New York: Dey St.: 258; and Kestenbaum, David. 2017, June 30. 'The Magic Show—Act One', *The American Life*. https://www.thisamericanlife.org/619/the-magic-show

49 For examples of such literature, see DePaulo, B.M., Kashy, D. A., Kirkendol, S. E., Wyer, M. M., and Epstein, J. A. 1996. 'Lying in Everyday Life', *Journal of Personality and Social Psychology*, 70: 979–995. https://doi.org/10.1037/0022-3514.70.5.979. and Newton, P., Reddy, V. and Bull, R. 2000. 'Children's Everyday Deception and

the sub-field of Symbolic Interactionism, social interaction is often conceived as entailing mutually monitored acts of self-presentation.[50] Herein, individuals:

- strive to control the image of themselves they express to others through what information they give and conceal through speech, dress, comport, facial expressions, etc.;
- attempt to uncover others' self-presentation performances on the basis of what others intentionally provide by way of information and what they inadvertently give away;
- recognize that others, in turn, are trying to uncover their self-presentation by what the individual intentionally gives and inadvertently gives away.[51]

Within such tangled cycles of presentation-discernment, complete honesty and forthrightness with one another can *threaten* our ability to get along harmoniously. In contrast, tactful words, discretion and other ways of maintaining polite fictions are commonplace means of avoiding overt conflict and preserving relations.[52] Such forms of pretense can become so deep that individuals no longer consciously strive to create an illusion for others. Instead, ways of acting become internalized and taken for granted.[53]

Within my sessions—that is to say, small group interactions between acquaintances—the potential for participants to engage in offering fabrications geared towards managing an impression of the scene and each other were ever-present. Therefore, I could hardly rule out deception directed towards me, whatever I heard from participants or read from their faces. But I could not definitively discern deception either. The same ways of speaking, gesturing and behaving that are used in face-to-face interactions to display honesty are those that accomplish subterfuge.

Performance on False-Belief Tasks', *The British Journal of Developmental Psychology*, 2: 297–317. https://doi.org/10.1348/026151000165706.
50 Scott, Susie. 2015. 'Intimate Deception in Everyday Life'. *Studies in Symbolic Interaction*, 39: 251–279. https://doi.org/10.1108/S0163-2396(2012)0000039011
51 Goffman, E. 1959. *The Presentation of Self in Everyday Life*. Harmondsworth: Penguin.
52 Also see Adler, J. 1997. 'Lying, Deceiving, or Falsely Implicating'. *The Journal of Philosophy*, 94(9): 435–452. https://doi.org/10.2307/2564617.
53 Hochschild, A.R., 2003. *The Managed Heart: Commercialization of Human Feeling*. London: University of California Press. https://doi.org/10.1525/9780520930414.

Reflecting on different theories of the mind—that is, how we understand the perspective and intentions of others—philosophers Shaun Gallagher and Dan Zahavi contended:

> In most intersubjective situations we have a direct understanding of another person's intentions because their intentions are explicitly expressed in their embodied actions and their expressive behaviors. This understanding does not require us to postulate or infer a belief or a desire hidden away in the other person's mind.[54]

Whether or not this is the case for most interactions, reflecting on my experiences provided many grounds for doubting these contentions relating to magic performances.

Chapter 2 included some of the paradoxical aspects of knowing the other that I experienced in learning to undertake self-working card magic instructions. This included how my growing experience with magic both brought me *closer to* and *away from* being able to appreciate the perspective of audiences. Likewise too, through performing magic for audiences, I developed a sense of the potential for magic as a method for making a *connection* with others. Yet my growing concern with deception in 2018 and 2019 brought concerns about my reason for *disconnection* from others. Both came together in a recognition that seems aptly labelled as bittersweet.

---

54 Gallagher, Shaun and Zahavi, Dan. 2007. *The Phenomenological Mind: An Introduction to Philosophy of Mind and Cognitive Science*. London: Routledge: 187.

4. Natural and Contrived

Borrowing a term from Donna Haraway, the previous two chapters depicted conjuring as an activity of sympoiesis or 'making-with'.[1] Making-with was accomplished through varied actions: mutual physical coordination; imagining others' perceptions; sharing attention on an object; forwarding the existence of definite rules, and so on. Yet, in important respects, the making-with as part of an activity of make-believe did not entail a sharing-between.[2] Magic is predicated on the possibility of fostering fundamentally dissimilar affective and perceptual experiences between conjurors and audiences.[3]

More than just noting aspects of sharing and divergence, the previous chapters elaborated ways in which learning magic entailed a paradoxical appreciation of doubleness: both becoming closer to and more distant from an understanding of others and self.

Putting these matters in theoretical terms, this doubleness could be glossed through its relation to 'the natural attitude'. Within the social sciences, Alfred Schutz's conception of the natural attitude has served as a starting orientation for understanding individuals' day-to-day experiences.[4] As Schutz argued, people habitually operate with important but often unrecognized assumptions: we perceive the world as it is; we experience it as others do; and our experiences can serve

---

1 Haraway, Donna. 2016. *Staying with the Trouble*. Durham, NC: Duke University Press: 5.
2 This is a long running theme in attempts to theorize magic, as in Blackstone, Harry. 1977. *Blackstone's Secrets of Magic*. North Hollywood, CA: Wilshire.
3 Those occasions when an effect is so persuasive that magicians themselves are amazed by what they experience can mark a high point in performance careers; for instance, see Kestenbaum, David. 2017, June 30. 'The Magic Show—Act Two', *The American Life*. https://www.thisamericanlife.org/619/the-magic-show/act-two-31; Granrose, John. 2021. *The Archetype of the Magician*. Agger: Eye Corner Press: 52–53; and Regal, David. 2021, February 9. *Bristol Society of Magic Lecture*.
4 Schutz, A. 1962. *Collected Papers* (Volume 1). The Hague: Martinus Nijhoff.

as a guide for gauging future actions. When such taken-for-granted presumptions come into doubt, then people routinely engage in forms of repair work (for instance, contending that someone erred) that restores a sense that there is a shared world known-in-common.

The past chapters have suggested ways in which learning magic entails an uneasy relation to the natural attitude. This did not take the form of out-and-out doubt about perceptions. My perceptions provided the precondition for practice. The confirmation of the mundane nature of the experience was a regular outcome of many hours of solitary rehearsal. However, learning magic offered many occasions for coming into doubt with how this world is 'seen in common, heard in common, felt in common and in these ways a world which is sensible in common'.[5] More than my mere personal experiences, professional magicians regularly question how they know themselves and others.

In continuing to approach magic as *deft contrariwise performance*, this chapter considers some of the entanglements between the umbrella notions of 'natural' and 'contrived' as part of instructional guidance. In seeking to manipulate the way others make sense of the world, conjurors labor to thwart detection of their secreted methods. A prime way they do so is by appearing to act naturally. As will be elaborated in this chapter, however, seeing naturalness and acting naturally are routinely treated as abilities that must be cultivated. How this cultivation is done is the focus of this chapter. Specifically, I turn to examine how those learning magic are taught to relate to people and objects in contrived ways that render their behavior natural, uninteresting, as expected and so on.

Acting Natural

> In all cases of palming the deck should be covered for the smallest possible space of time, and the covering and exposing should be made under some natural pretext, such as squaring up the cards, or passing the deck to the other hand, or changing its position in the hand, or turning it over.[6]
>
> – S.W. Erdanese (pseudonym)

---

5 Girton, George D. 1986 'Kung Fu'. In: *Ethnomethodological Studies of Work*, H. Garfinkel (Ed.). London: Routledge & Kegan Paul: 70.
6 Erdnase, S.W. 1955. *The Expert at the Card Table*. Mineola, NY: Dover: 90–91.

4. Natural and Contrived

Just as control is a frequent theme when magicians reflect on their art, so too with naturalness. The quote above from Erdanese's classic book *The Expert at the Card Table* illustrates how being natural is regarded as necessary and advantageous. Acting in this way provides cover for undertaking sleights—in this case, 'palming' or concealing the cards in one's hand. In wishing to make contrived actions appear otherwise, striving to be natural is a frequent goal.

Doing what is natural is often portrayed as a complicated business by conjurors. For instance, as with *The Expert at the Card Table*, Hugard and Braué's *The Royal Road to Card Magic* makes recurrent reference to naturalness. As one example, Hugard and Braué stipulate that:

> Practically everyone, when beginning to practise the palming of cards, will be careful to keep the fingers curved naturally but will overlook the importance of having the thumb lie in its natural position along the side of the hand. When the thumb extends at a right angle from the hand, a reflex action which must be overcome, its unnatural appearance at once attracts attention to the hand and arouses suspicion.[7]

Herein, the mention of 'natural' in the first sentence refers to action that must be deliberately worked at. While lodging a card between one's fingers and the ball of the thumb, the fingers must be made to appear relaxed, as they would be in the absence of the palmed card. In contrast, while thumbs generally lie along the sides of hands, this is not always so during the peculiarities of handling palmed cards. Instead, the habitual response of the thumb needs to be counteracted in order for it to assume the desired natural-appearing position.

With the frequency and significance attached to naturalness in entertainment magic, it is perhaps not surprising it is subject to alternative portrayals. Elsewhere, in explaining how to glimpse a specific card secretly, Hugard and Braué note:

> We shall suppose that you have handed the deck to a spectator to be shuffled. When he has done that, hold out your right hand to take back the deck, purposely holding it rather high so that he will have to raise his hand to give you the deck. Take the pack with your thumb underneath it on the face card, your fingers on the back. At that moment it is natural

---

[7] Hugard, Jean and Braué, Frederick. 2015. *The Royal Road to Card Magic* (Video Edition). London: Foulsham: 87.

for you to glance at the cards, and by tilting them ever so little with the thumb you can glimpse the index of the bottom card at the inner left corner.[8]

In this description, naturalness is positioned as built into the details of the coordinated interaction. And yet, subsequently they go on to state, 'Get the glimpse and then look at the spectator. Make the action a natural one, and no one will have the least suspicion that you have seen the bottom card.'[9] This later sentence, however, suggests the naturalness of the glimpse has to be secured above and beyond the barebones of the coordinated interaction. Wider than this specific example, within *The Royal Road to Card Magic*, 'natural' varies between being understood as a universally shared interpretation of an action ('a natural way of squaring the cards'[10]) and as a characterization dependent on the alignment of action with the specific presentational style of individuals.[11]

Against his experiences of the question-begging meaning of evocations to be 'natural', magician Darwin Ortiz considered, and then rejected, two ways of settling what is natural. One was based on community standards and the other on analogic resemblance. For him, naturalness should not be determined by what the majority of magicians regard as such. Neither is it a matter of mimicking how lay audience members would behave if they were to handle rings, coins or other props for themselves. Naturalness instead derives from actions being *justified* (the motivations for them within the presentation are immediately apparent to audiences) and *economical* (they are done with what is perceived to be the least amount of energy). Since acting in ways that secure both judgements in the minds of audiences takes considered effort, 'naturalness doesn't come naturally'.[12]

Just as what counts as natural has been subject to different interpretations, so too has the place of naturalness. Hugard and Braué aimed to hold a 'mirror to nature' through their art and therefore

---

8   Ibid.: 67.
9   Ibid.
10 Ibid: 190.
11 As in Hugard, Jean and Braué, Frederick. 2015. *The Royal Road to Card Magic* (Video Edition). London: Foulsham: 89.
12 Ortiz, Darwin. 1994. *Strong Magic*. Washington, DC: Kaufman & Co.: 316. For a wider discussion on the need for justified movements in the case of acting, see Lecoq, Jacques. 2000. *The Moving Body*. London: Bloomsbury.

sought to undertake natural and (as much as possible) relaxed forms of card handling. In doing so they operated within a style referred to as *modern magic*, a style that still heavily influences the standards by which conjurors are judged today.[13]

As Wally Smith has detailed, the origins of modern magic as a performance style can be seen as a response to the type of stage shows prominent up to the mid-19th century.[14] The latter type entailed the use of elaborate stage set-ups, dim lighting, numerous props and other features that signalled to audiences that relevant features of the scene were being obscured from view. Even if audiences could not identify how the bagginess of the magician's robe mattered, they could surmise that it did. In contrast, the modern stage magic that came to the fore in the mid-19th century strove for minimally visible objects that were often depicted as incidental. The overall intention was to make audiences 'confident that they had seen all they needed to see'.[15] In this respect, it can be said that what was absent from performances was as important as what was present. Along with this orientation to their surroundings, the personas of magicians shifted from the exotic and mystical to the commonplace and conventional. During the 1920s and 1930s, the widespread movement in magic from the stage with large props into small group interactions with everyday objects—what would become labelled as close-up magic—provided a further realization of the modern style. Within close-up magic, a commonplace objective is rendering techniques such as card sleights either unseen or seen but unnoticed.

As Smith identified, at the heart of this style remains a basic tension: its desire for the appearance of effortlessness is based on carefully designed mechanical repetition and minutely choreographed action. As a result, one danger conjurors face is that as audiences become more familiar with magic, it becomes harder to pass off carefully crafted movements as incidental. Another danger that can be proposed centers on the display of skill. Whereas overt displays of skill hazard reducing magic to something similar to juggling, the absence of such displays risks encouraging audiences to believe anyone could perform the feat.

---

13 See Tibbs, G. 2013. 'Lennart Green and the Modern Drama of Sleight of Hand', *Journal of Performance Magic*, 1(1). https://doi.org/10.5920/jpm.2013.1119.

14 Smith, Wally. 2015. 'Technologies of Stage Magic', *Social Studies of Science* (June), 45: 319–343. https://doi.org/10.1177/0306312715577461.

15 *Ibid.*: 325.

The demands on magicians striving to deceive through naturalness share similarities to other situations in which individuals try to pass themselves off. Garfinkel's classic study of Agnes, a 19-year-old transsexual that sought a sex change, details the incessant measures that can be necessary in order to continuously accomplish being taken by others as a woman when socialized as a man. In the case of Agnes, her studious efforts to conform to a highly conventional image of what it meant to be a woman not only included how she spoke, comported herself, expressed opinions and so on, but how she accounted to others for how she spoke, comported herself, expressed opinions and so on. Accounting-for included coming up with various excuses as to why she had to refrain from certain activities (for instance, not being in the mood for swimming or being modest in situations that required nudity). Agnes' case illustrates the considerable but often overlooked efforts necessary to be taken as acting naturally according to cultural expectations of the day.[16] As with Agnes, magicians need to attend to how their gestures, patter, gaze and so on aligns with the range of commonplace cultural expectations for how magicians ought to behave.

In general terms, magicians in performance situations endeavoring to be natural face both fewer and more demands than Agnes. Fewer because, while Agnes constantly feared any slip up might betray her carefully constructed identity as a conventional woman, a conjuror typically needs a natural style to cover for specific critical moments of deception. More because, unlike those Agnes encountered, audiences to a magic show anticipate a conjuror is trying to deceive them, even as the conjuror attempts to persuade audiences that they have seen all they need to see.

Although hardly unique to magic, conjurors need to attend to how their ongoing behaviour establishes expectations for what counts as consistent and thereby justified conduct (for instance, regarding rhythm and speed, overall demeanour). Any sensed deviations from the emerging pattern performers construct through their individual

---

16   For a wider analysis of these points in relation to gender generally, see Butler, J. 2007. *Gender Trouble: Feminism and the Subversion of Identity*. London: Routledge Classics. For a wider analysis specific to magic see Beckman, Karen. 2003. *Vanishing Women: Magic, Film, Feminism*. Durham, NC: Duke University Press. https://doi.org/10.1215/9780822384373.

style are liable to raise suspicions in audiences.[17] 'Sensed' is an important qualification. Whether or not those suspicions end up being well-grounded, their arousal can detract away from the prospects of generating wonder.

And yet, despite all of these points, it is the very anticipation for deception in magic that provides a basis for realizing it. As a simple example, to demonstrate to audience members that a deck of playing cards is standard, not sequenced in a particular order and so on, a magician can spread the cards face up on a table. While doing so can serve to reassure the audience, it can also be the very technique by which the conjuror identifies the position of critical cards. As Smith contended:

> magicians go to elaborate lengths to make their actions *accountable* to spectators, those present and those watching on television [...] And the audience patiently accepts this pattern of making things accountable, because they are involved in a live social interaction in which it is part of the job of the magician to assure them that everything is fair and above board. It is the very need for the magician to be accountable, and the acceptance by the audience that such acts of accountability are ordinary in themselves, that gives the magician space to conceal secret actions...[18]

As such, the understandable and intelligible appearance of the magicians' actions serve both to make those actions sensible as well as to dissimulate. Manipulative displays of accountability need not be limited to the actions and objects of magicians. The working environment of performances can be marshalled to persuade the audience that everything is fair and above board. Filming magic 'on the street' with seemingly random members of the public is a way to diminish the suspicions that audiences attach to set piece stage shows.[19]

And yet, as with so many other types of naturalized deceptions, the positioning of paraphernalia, actions and settings is a delicate operation. Drawing too much attention to the ordinariness of what is before an audience might in itself generate niggling suspicion and sobering disengagement where there was once unquestioned acceptance and

---

17   Fitzkee, Dariel. 1945. *Magic by Misdirection*. Provo, UT: Magic Book Productions.
18   As elaborated in Smith, Wally. 2016, April 8. 'Revelations and Concealments in Conjuring'. *Presentation to Revelations Workshop*. Vadstena.
19   Turner, E. 2016. '"I Am Alive in Here": Liveness, Mediation and the Staged Real of David Blaine's Body', *Journal of Performance Magic*, 4(1). https://doi.org/10.5920/jpm.2016.03

lively participation. As a result, some instructions call for a minimal approach in which accountability is secured through indirect means.[20] For instance, rather than verbally justifying why someone is searching their pockets ('Where is my pen? I have forgotten…Oh, here it is.'), the recommendation is simply to produce a pen casually from a pocket and then make use of it. Accountability becomes 'self-evident'.[21] As still a further 'and yet', at times a conjuror might seek to generate suspicion and disengagement at specific points to thereby direct attention this way and that too.

Thus, being natural is a subtle undertaking.

It can also be a contrived one. As an example, at the start of the 20th century the American magician William Robinson achieved professional acclaim in a modern style through adopting the stage presence of a Chinese magician called Chung Ling Soo. As Chinn argues, Robinson did not persuasively present himself as authentically Chinese because his outward appearance and conduct were flawless. Instead, he was able to pass as Chinese through the *defects* of his performance. Race relations at the time in which Chinese people were stereotyped and treated as culturally 'other' from white audiences combined with the artifice of stage magic to produce a situation in which 'more unbelievable is not only better, but paradoxically believable'.[22] Not for the first or last time in history, the case of Chung Ling Soo illustrates how marketable portrayals of authenticity depend on performers playing to cultural beliefs about the exotic and familiar, as well as on audiences' willingness to go along with impersonations.[23]

The points above about naturalness in modern magic have applied to situations in which an identifiable performer acts on a literal or figurative stage. As the audience is anticipating deception, those on stage need to hide certain actions, even as they work to convince others that there is no attempt to do so. However, offstage, many individuals draw on techniques in conjuring to produce feelings of astonishment

---

20 See Fitzkee, Dariel. 1945. *Magic by Misdirection*. Provo, UT: Magic Book Productions.
21 Earl, Benjamin. n.d. *Real Deck Switches*. Sacramento, CA: Vanishing Inc.
22 Chinn, Mielin 2019. 'Race Magic and the Yellow Peril', *The Journal of Aesthetics and Art Criticism*, 77(4): 425.
23 Rosenthal, Caroline. 2021. 'The Desire to Believe and Belong'. In: *The Imposter as Social Theory* Steve Woolgar, Else Vogel, David Moats, and Claes-Fredrick Helgesson (Eds). Bristol: Bristol University Press: 31–52. ttps://doi.org/10.1332/policypress/9781529213072.003.0001.

in ways that blur, if not downright dissolve, the line between staged performances and everyday interactions. For instance, consider Jay Sankey's 'Unreal Experiments'. As Charles Rolfe has examined, these experiments entailed dispensing with an identifiable magician-figure, a self-recognized witnessing audience and a planned outcome.[24] In one such experiment, Sankey posed as a customer buying sweets that did not have enough money. When the cashier insisted on payment, he opened one of the sweet packets to find a 20-dollar bill inside.

In a related vein, the magician blogger working under the name *The Jerx* advocates an 'amateur' style in which befuddling outcomes that defy everyday expectations are realized in informal social interactions in which no one overtly 'performs' for another. As an example of this style, he recounted how:

> I'd be out to eat with someone and while we talked and waited for food I'd be making little tears in a business card [...] When I was done I'd just set down two linked paper circles torn from one business card and never say a word about it. When the person noticed it my response was, "Huh?" That's always my first response with the distracted artist presentation. "Huh? Oh.... that's bizarre." You've got to slow-play this. You're not being humble and you're not acting like you're not responsible for it, you're just not taking credit for it because you don't see it as something to take credit for. These things just happen.[25]

In such ways, this kind of performance of non-performing sets out to harness a sense of the naturalness of actions (for instance, likening them to doodling), combined with a recognizably odd outcome (for instance, the two linked paper circles) with a portrayed need *not* to be accountable within an interaction in order to bring about a feeling of strangeness. As with the pranks of the past television program *Candid Camera*, or the high jinks in the more recent *The Carbonaro Effect*, Sankey and *The Jerx* set out to discover what response is provoked rather than to realize a specific one. A challenge in doing so is being proficient enough at undertaking the required actions that they are not regarded as planned.[26]

---

24 Charles, Rolfe. 2020. 'Theatrical Magic and the Agenda to Enchant the World', *Social & Cultural Geography*, 17(4): 574–596. https://doi.org/10.1080/14649365.2015.1112025.

25 The Jerx. 2015, June 12. 'Presentation Week Part 5: The Distracted Artist Presentation', *The Jerx*.

26 Regal, David. 2019. *Interpreting Magic*. Blue Bike Productions: 66.

However, the lack of acknowledgement or refutation that any trick (as such) is being performed also means that the immediate audience can find reasons to dismiss the idea that anything magical or noteworthy has taken place. In this way, *The Carbonaro Effect*, Sankey and *The Jerx* occupy an ambiguous relation to entertainment magic, even as they directly draw on its methods.[27]

The Mirror of Nature

In light of the previous analysis of the negotiated meaning and place of naturalness, the next sections attend to how those teaching magic strive to instil naturality.

I do so by first considering the situated practices associated with one simple visual aid for gauging the naturalness of actions: a mirror. Practicing in front of a mirror is a common technique for conjurors, and it was one that I utilized once I started practicing sleights. The advantages of this technique, compared to imagining how you appear, can be pronounced. For instance, a common sleight used in conjuring is the 'double lift'—lifting two cards from the top of the deck as one. At times, however, ensuring two cards (and only two cards) are lifted (and together) can be a fiddly operation, with the thumb and fingers being varyingly maneuvered. When this kind of fiddling occurs, as it has done many times for me, it is easy to imagine that the card manipulation would be blatant to onlookers. Through enabling simultaneous feedback between actions and one's own evaluation of them, practicing in front of a mirror visually confirms that a slightly angled deck affords considerable scope for occulting hand movements. With this cover, what otherwise might well appear erratic and suspicious becomes fluid and unremarkable. Although I have practiced this lift hundreds of times to make the handling appear effortless, again and again, I still feel compelled to do so in front of a mirror to confirm the sleight is not readily detectable.

In this respect, then, mirror practice brings together the possibility of perceiving yourself as another person would, with the underlying

---

27 For a further related discussion concerning what has been called 'bizarre' magic, see Taylor, N. 2018. 'Magic and Broken Knowledge;' *Journal of Performance Magic*, 5(1). https://doi.org/10.5920/jpm.2018.03

knowledge and experiences of a magician. As such, the mirror functions as a relational device in self-other relations. Psychologist Jacques Lacan, for instance, famously claimed specular images secure self-fantasies: against internal tumultuous bodily drives and fragmentary thoughts, one's mirror image serves as an aspirational representation of a stable, whole and coherent self.[28] In certain respects, practicing magic entails a kind of identification with the mirror image. Whatever one's fraught internal affective and physical states, as a representation of self that is observable to others, the mirror image serves as a touchstone for gauging naturalness.

Or at least in certain respects it does. Generally, more than just providing an external reflection, mirrors extend an invitation for internal reflection.[29] In my case, the very need to revisit the double lift in front of a mirror, for instance, signals the lack of definitiveness of specular self-witnessing. One limitation is that the seeing done during mirror practice is the situated seeing of one who knows what to look for. Yet, audiences may not perceive or give significance to what might be glaringly obvious to a magician.[30] Indeed, as magicians operating in the modern style often (but not always) strive to *not* bring attention to card manipulations in the first place, what might be at the center of concern for a magician looking in the mirror may not be meant to be relevant to audiences at all. Looking in a mirror—properly—thus needs to involve something else than looking closely. How and where to gaze, though, are not necessarily straightforward.

Conversely, though, while audiences might not apprehend card sleights, a common refrain in professional advice is they can be highly perceptive in sensing certain kinds of subtleties. When conjurers unconsciously tense up in anticipation of a difficult manipulation, the audience can pick up on it 'like a dog senses fear'.[31] As a response to both sets of considerations, instructions often direct learners not to

---

28 Lacan, J. 1977. *Ecrits: A Selection* (trans. Alan Sheridan). New York: Norton.
29 For a discussion of both possibilities, see Colie, Rosalie. 1966. *Paradoxia Epidemica: The Renaissance Tradition of Paradox*. Princeton, NJ: Princeton University Press.
30 Along these lines, in teaching situations magicians often tell their students to ignore what they are seeing because they know what to look for, see 1:17:20 in DaOrtiz, Dani. 2017. *Penguin Dani DaOrtiz LIVE ACT. Available from:* https://www.penguinmagic.com/p/11142
31 Regal, David. 2019. *Interpreting Magic*. Blue Bike Productions: 143.

get preoccupied with making sleights visually perfect, but instead to engage with audiences and thereby produce a relaxed atmosphere that disarms scrutiny.[32]

As a result of the competing considerations noted in the previous paragraphs, learning through mirror practice entails a delicate process of seeking to refine the physical handling of cards through greater visual discrimination, while also attending to whether and how 'seeing' is relevant.

Performing in a Material World

This analysis of mirror practice also offers an opportunity for a slight diversion to consider magic as a socio-material activity. One way to conceive of this, from the perspective of the magician, is as a relation of asymmetrical dualism. Just as it has been said that conjuring is always about 'control, control, CONTROL!'[33] of their audiences, much the same could be said of the material world. The job of the magician is to render things and environments into docile and manageable objects so that they can serve as instruments of performance.

Striving for this sort of mastery has obvious rationales within attempts to accomplish planned effects. Yet, whether the aim of control is a good descriptor for characterizing the practice of conjuring or any other activity is another matter. As Andy Pickering has contended, the notion that an entity in the world can be rendered into a passive object to be manipulated at will requires it 'does exactly what we intend *and no more*'.[34] However, 'even when we successfully arrive at a situation in which the world does our bidding, it typically does something else *as well*.'[35]

The case of the development of bubble chambers in particle physics is one of the many examples Pickering has given to illustrate how that 'something else' comes about and what it can entail.[36] In the case of the

---

32 For instance, see https://www.youtube.com/watch?v=NUrtygFXPDQ
33 Thun, Helge 2019. 'Control', *Genii*, December: 71.
34 Pickering, Andrew. 2017. 'In Our Place: Performance, Dualism, and Islands of Stability', *Common Knowledge*, 23(3): 392. https://doi.org/10.1215/0961754x-3987761. Emphasis in original.
35 Ibid. Emphasis in original.
36 Pickering, Andrew. 1995. *The Mangle of Practice: Time, Agency, and Science*. Chicago: University of Chicago Press: Chapter 2.

inventor of the bubble chamber, Donald Glaser, his attempts at devising novel forms of manipulating atoms in the 1950s consisted of back-and-forth movement between human and non-human agency. As Pickering set out:

> Over a couple of years, Glaser, as an active human agent, would put together some configuration of apparatus. Then, while the agency of the material object took over, he became passive, standing back with a movie camera in his hand to record what his latest setup would do. Switching back to an active role, Glaser would then react to the machine's performance (which was usually not what he wanted), and the dance would continue. Its upshot was a new instrument that did new things in a new way—revealing the paths of elementary particles as strings of bubbles forming in a liquid (and winning Glaser a Nobel Prize). At the same time, Glaser himself was transformed: shifting from small science to big science, becoming the leader of a sizable group and becoming famous, changing his ideas about how bubbles form and what bubble chambers should look like, and moving from one subfield (cosmic-ray physics) to another (accelerator-based experimentation).[37]

This example illustrates how attempts at controlling the world can result in an alteration of the person notionally in command. Parallel cases of complex, iterative adaptation can be found in the case of conjuring. Today, with the rise of social media as a platform for performing magic, leading innovators in this medium have contended that the need for immediacy in effects is radically shifting the compulsory skills of magicians and previously held community ideals for what it means to act in an apposite manner.[38]

How I have characterized learning magic in this book—as a sociomaterial practice entailing paradoxical movements between becoming closer to and more distant from an understanding of others and self—is aligned with treating agency and control as highly negotiated. My experience entailed interactions with the world, in which my agency and my abilities have shifted over time through my attempts to exercise agency and perceive the world.

---

37 Pickering, Andrew. 2017. 'In Our Place: Performance, Dualism, and Islands of Stability', *Common Knowledge*, 23(3): 382–833. https://doi.org/10.1215/0961754x-3987761.

38 For instance, Elderfield, Tom. 2019, June 17. *The Insider*. https://www.vanishingincmagic.com/blog/the-insider-tom-elderfield.

Experienced magicians, too, have recognized ways in which engaging with the material world can change them. For instance, despite mirror practice being a common technique, the authors of *The Royal Road to Card Magic* advise against such training. As Hugard and Braué warn, 'mirror watching has tendency to cause the eyes to widen; this isn't attractive and can become a fixed habit'.[39] Herein, rather than magicians arranging mirrors, mirrors arrange magicians. They do so by conditioning viewers into unnatural forms of acting, and in ways that might not even be noticed by the conjuror. Ben Earl too has cautioned against the extensive use of mirrors because such practice can lock magicians into rigid movements and positions that appear persuasive within the confines of solo mirror practice, but are not so within the dynamism of live performances.[40] Perhaps even more invasive, Augusto Corrieri contended the extent of his mirror practice has resulted in a dissociation: rather than seeing through his own lived senses, once on stage he cannot but see himself from the outside, as if he is an audience member.[41] For such reasons, a neat subject-object distinction become untenable. Magicians configure mirrors to suit their purposes, but the capacities of magicians emerge through their mirror-use.

Finding Naturality

While, in the abstract, acting naturally ought to come easy, the previous sections surveyed some of the many questions, choices and implications associated with being seen to be behaving naturally in conjuring. In doing so, I wished to underscore its cultivated and contrived status. Following on from the previous sections, this one further attends to the achievement of naturalness by examining how this appearance is taught. Given the previous argument, one question that can be asked of instructional training is how it reconciles treating naturalness as both readily apparent to any onlooker and not adequately appreciated by learners.

---

39 Hugard, Jean and Braué, Frederick. 2015. *The Royal Road to Card Magic* (Video Edition). London: Foulsham: 219.
40 Earl, Ben. 2020. Deep Magic Seminar, 18 July.
41 Corrieri, Augusto. 2016. 'An Autobiography of Hands', *Theatre, Dance and Performance Training*, 7(2): 283–229. https://doi.org/10.1080/19443927.2016.1175501.

Two instructional resources that figured in my learning are examined: an audiovisual DVD by Ben Earl and a face-to-face masterclass by Dani DaOrtiz. I became aware of both through their lectures at the first magic convention I attended: the 2019 Session Convention organized by Vanishing Inc. The lasting impressions their performance made on me led me to seek them out afterwards. For both instructional resources, I consider not only the prominent place given to naturalness but how the acclaimed performers varyingly make naturalness witnessable (or not) to learners. As will be suggested, how this is done is a nuanced sociomaterial, undertaken wherein students are both encouraged to see what is plainly in front of them and cautioned against taking what they see at face value.

Switching Decks, Switching Registers

Benjamin Earl's DVD titled *Real Deck Switches* provides instructions for variations of 25 ways to covertly change one card deck for another during a show. As with many audiovisual tutorials, *Real Deck Switches* positions the learner as an audience. The primary camera offers a view of Earl as if positioned straight across a playing table.[42] Through this set-up, viewers are invited to see for themselves—again and again—how convincing the deck switches appear. And yet, while switches are demonstrated, what is put on display is also acknowledged as implicating a sense of what is not shown. For example, at various points, Earl speaks to how the switches would be seen from angles not displayed by the camera frames. Instead of seeing for yourself, in these instances, the learner is asked to imagine what the deck handling would look like. In other ways, too, what is visibly displayed points toward what is not shown. On occasions, card handling that would normally take place under the table is displayed on top of the table. On other occasions, Earl's handling is obscured by the table. At these latter points, audiences need to envisage what is taking place out of sight.

Concerning naturalness specifically, similar alternations in what is offered to the learner can be identified. In line with the style of modern magic, in *Real Deck Switches* Earl repeatedly makes the case for the

---

42 A set-up that has been the source of some criticism; see https://www.themagiccafe.com/forums/viewtopic.php?topic=659980&forum=218

importance of achieving a smooth, economical and relaxed appearance. At times, the presence of such qualities in the card handling is treated as available to viewers by simply seeing his displayed movements. Switches are accompanied by verbal evaluations so that the movements appear natural, but without any further pointers as to why. As a result, a central element of the work for learners in comprehending the instructions is spotting for themselves the qualities of the card handling that exemplify naturalness.

At other times though the situation is presented as more complicated. One way this is done is by distinguishing between *feeling* and *looking*. For instance, Earl comments:

> The thing with these switches is to try to make them feel right. As well as look right, it is much more the feeling of the cards being cut than anything else. If you can get the timing and the movements soft, then they tend to feel deceptive and that is really the point.[43]

In this advice, how actions 'feel' is said to be more important than how they 'look'.

Elsewhere in *Real Deck Switches*, feelings are said to be able to override what is logically derived and visually witnessed. For instance, in working through a switch in which a discrepancy exists between how one deck leaves the table and another is brought in, Earl argues:

> Now it kind of doesn't make much sense. Because it is like the ((pause)) deck has morphed through the box. ((switch performed)) It should be on the other side. But how it feels ((switch performed)). To me, it just feels like you are putting the box away and the deck is still there. It does not really feel like the deck goes out of view.[44]

As such, learners are encouraged to harness a sense of a feeling/impression that cannot be reduced to the physical positioning of cards, the card box and his hands. This encouragement is given even as learners' attention is directed toward visually scrutinizing these precise positionings. The figurative pointing out of the discrepancy is thus no

---

43 Emphasis in delivery.
44 Emphasis in delivery. As an additional point to underscore, in this quote, how a move feels is not presented as definite and assured, but as a matter of individual opinion. Elsewhere, Earl likewise appeals to his own judgement by stating 'It just, for me, *feels* right. So you are in this position here. We are going to look for the box. Here we go, let's open that, and it really does not feel like a switch happened.'

simple undertaking, but a construing of the objects and movements that demands active involvement of the learner to make sense of what is being instructed by the instructions.[45]

Following such instructions necessitates trying to appreciate what is being referred to by this feeling/impression and to recognize how the switch possesses this quality. The way I created a feeling/impression was to refrain from a pinpoint focus on the deck, and instead to pan back my visual field as if I was taking in the scene as a whole. With this kind of orientation, even though I knew there was a discrepancy, I did not register a visual jarring associated with the placements of the decks. Through doing so, my way of looking conditioned my generated feelings.

At other points in *Real Deck Switches*, however, learners are instructed to not rely on feel, but instead to rely on looks. In attempting to palm a whole deck, for example, Earl notes that the deck certainly can feel awkward in the hand. Rather than relying on that feeling as a guide to what will be perceived by audiences, his suggestion is to attend to what is visibly displayed. He encourages this by demonstrating for the camera how palming a whole deck can be undetectable through careful positioning of the deck in relation to the table and his body.

Although distinctions are made on numerous occasions between feeling and looking, nowhere in *Real Deck Switches* are set definitions given for the two. At times, feeling is explicitly related to the touch of the magician. For instance, Earl suggests the decks should feel loose, soft and light in the hands. At other times, though, what is felt by the magician is contrasted with what is perceived by the audience.

As a result of the various kinds of positioning noted above, a vital element of work associated with following the instructions of *Real Deck Switches* is shifting between assessing how the imperative to feel relates to both the embodied experiences of the magician and the audience.

---

45 For a wider discussion on how figurative and literal pointing can entail complex inter-relation between participants, see Goodwin, C. 2003. 'Pointing as Situated Practice'. In: *Pointing: Where Language, Culture and Cognition Meet*, S. Kita (Ed.). Mahwah, N.J.: Lawrence Erlbaum; and Kendon, A. 1986. 'Some Reasons for Studying Gesture', *Semiotica, 62*: 3–28.

Naturalness Through Sensation

The second set of instructions I want to consider are those given in a face-to-face masterclass with the world-renowned magician Dani DaOrtiz. I attended one of these classes over 26–28 July 2019 at his villa outside of Malaga, Spain. Within a setting that entail 'prestigious imitation',[46] students in masterclasses become socialized and skilled into a 'community of practice' through attending to the language and behavior of a recognized authority figure.[47]

Widely praised for a casual yet chaotic manner of card handling, in which exacting control appears unthinkable, DaOrtiz offers a highly contrasting presentational style to that of the smooth and soft precision of Earl. Despite this, cultivating feelings and perspectives to achieve a sense of naturalness is likewise central to DaOrtiz's approach.

While no dictionary-type definition was given for 'natural' during the weekend, DaOrtiz repeatedly used the term to signal how the magician's actions should be expected, justified and regarded as ordinary from the spectator's point of view. Within our small group sessions, DaOrtiz outlined his preparation strategy for achieving this impression of naturality. As he explained, naturalness was not realized by an attitude of indifference to planning, but rather the contrary. As conveyed, his shows were studiously prepared for and his interactions with audiences meticulously choreographed despite their haphazard appearance. Throughout the class, he offered step-by-step explanations for how to manipulate cards, how to position one's body and surrounding objects, as well as how to direct attention through glances and gestures in ways that would be taken as natural by audiences. For instance, we were taught how to position our feet on the ground when facing a spectator, so that the audience would expect from the subtleties of our contortions that we would subsequently turn back. This turning back provided cover for a sleight-of-hand manipulation. Justifications could work backwards in time too; actions were undertaken at one point to make what he had done previously look natural in retrospect.

Through such a rehearsed self-presentation, magicians' actions could appear uncontrived; indeed, in the chaotic style of Dani DaOrtiz,

---

46 Mauss, M. 1973 [1934]. 'Techniques of the Body', *Economy and Society*, 2(1): 70–88.
47 Wenger, E. 1999. *Communities of Practice: Learning, Meaning, and Identity*. Cambridge: Cambridge University Press.

the actions appear as out of control. Much of the instruction over the weekend consisted of him teaching us how to perceive magic from a spectator's point of view, so as to appreciate how such perceptions could be harness to produce wonder.

Chapter 5 examines some of the ways perceiving as a spectator was presented as more or less possible within the masterclass. In the remainder of this chapter, I want to concentrate on one specific aspect of perceiving as another: gauging the feelings of spectators. Spectators' feelings are at the center of DaOrtiz's instructions, because skillfully affecting them is taken as his end aim of magic, not skillfully handling playing cards. As he advocated:

> It doesn't matter if you do a palm, a double lift, or a shuffle, it does not matter. The final is what the spectator feels. And this is the effect. Not what you do here. ((spreads out an imaginary deck in his hands)) The effect is not the card in the bottle. The effect is the card inside an object near to me when the magician is there. ((points away)) S!#t ((opens arms to the side)) This is the effect. Not one signed card is inside a bottle. No, no. This is the physical effect. Wow is the effect you remember. And for that we need to work.

Feelings and naturalness were presented as intertwined because: (i) emotional engagement reduces the extent of critical scrutiny by audiences, and (ii) audience judgements that the magician's actions are expected, justified and ordinary heighten the resulting emotional reactions.

In terms of (i), DaOrtiz advocated a chaotic means of card handling, both because it fosters visceral sensations, and because it reduces the likelihood that spectators will attempt to (as well as successfully be able to) reconstruct the methods for effects.[48] Spectators cannot understand what happened if the magician acts naturally, because rational memory reconstructs the past using actions regarded as noteworthy. To act as expected therefore provides little by the way of mental hooks for recall.

---

48 To be clear, displayed chaos is located in the means, not the ends. This is to say that, while the process of handling cards should appear disordered, the end effect should be clear for spectators to grasp. Thus, whilst producing a sense of chaos, we were also instructed to construct moments of significance in our performances (so-called 'memory pillars') that could provide a straightforward basis for spectators to fashion a sense of what had taken place, and thereby the significance of a trick's culmination.

Instead, what lingers for spectators are the potent sensations associated with the culmination of a trick.

In terms of (ii), we were instructed how naturalness was achieved by doing and saying *less*. Rather than telling spectators, 'Put your card back here' whilst pointing to a deck, we were advised instead to simply say 'Put back' with our bodily and hand positioning providing cues about the precise placement. Through combining the shortened directives with a relaxed style, DaOrtiz contended, spectators wouldn't *think* or (more importantly) *feel* to question why their card was being put in that specific location. This lack of regard for the location of the placement would then underpin the generation of an emotional and sensational response to the subsequent identification of a chosen card.

Consistent with the importance attached to feelings, again and again in the masterclass, DaOrtiz sought to engender emotions and sensations within us as students by performing tricks in line with his style, by contrasting different presentations for individual effects, by play-acting as a naïve spectator, and by asking the students to act out scenarios. Through such enactments, what was sought was a kind of 'hot' authentication based on experiencing affective states, rather than the 'cool' authentication gained by hearing expert pronunciations alone.[49] That hot authentication was a bodily experience, routinely demonstrated through non-verbal actions (glosses of some of the non-verbal actions are listed in the right-hand column with the left-hand column providing a direct transcript):

Direct transcript	Non-verbal actions
Always the finale need to be a sensitive	*Takes hand to heart*
thing, never a rational thing.	*Points finger to his head*
Always sensitive. It's like a s#!t	*Slams his hand on the table.*
No	*Moves eyes and fingers in a gesture of thinking*

---

[49] To borrow a distinction from Cohen, Erik and Cohen, Scott A. 2012. 'Authentication: Hot and Cold' *Annals of Tourism Research*, 39(3): 1294–1314. https://doi.org/10.1016/j.annals.2012.03.004. See as well Ruhleder, K. and Stoltzfus, F. 2000. 'The Etiquette of the Masterclass', *Mind, Culture and Activity*, 7(3): 186–196. https://doi.org/10.1207/s15327884mca0703_06.

Direct transcript	Non-verbal actions
No, no, no, it's a look	*Leans forward and looks down*
and feel.	*Takes hand to heart*
Look	*Leans forward*
and feel.	*Replaces hand to heart*
Always	

In this instance, the hand gestures, talk, gaze and bodily movements undertaken served to underscore the divisions central to DaOrtiz's teaching in general and his speech at that movement—looking/feeling and rational/sensational. Recognition of those divisions and the importance of attending to feelings and sensations were repeatedly presented as vital by DaOrtiz. It was through these undertakings that magicians can generate feelings of wonder in others.

With the masterclass, DaOrtiz evoked feelings to assess different ways of performing well-known sleights. Doing so meant challenging how we as students might well have made sense of our experiences up until that point in time. For instance, as mentioned above, a double lift involves taking two cards from a deck together, typically to convince the audience only one card is in play. In speaking to the naturalness of moments using the double lift, DaOrtiz contended:

Direct transcript	Non-verbal actions
The double lift is one of the best	*Hand gesture pointing away*
technique; at the same time it is one of the worst technique.	*Hand gesture pointing toward chest*
Why?	*Spreads hands*
Because technically, physically, it is	*Rubs fingers together*
unbelievable. Sensationally, it is very bad.	*Clasps heart*

Direct transcript	Non-verbal actions
Look. I show you six of clubs. Yes?	*Overturns cards on the top of a deck in his hand thereby performing a double lift. Flips over cards again to be face down on deck.*
I put the six of clubs on the table.	
I snap my finger and change.	*Card on top of deck put face down on table*
And you say 'Wow'. It is very good. *!#%. But you are feeling two cards. Why? Because the *first* card you see here, not there.	*Finger snap. The card on table overturned to show the queen of spades.*
You don't know how I changed but you are feeling more than one card. Then if you are making sensations, this is what you need to do.	*Points to deck in hand and then to table.*
((Dani explains how to execute the double lift in line with his presentational style))	

Herein, through verbal and non-verbal actions, DaOrtiz seeks to bring to the fore an unease that he suggests audiences will experience in witnessing sleight of hand. It is a feeling of unease that is characterized as operating below the initial positive sensation ('Wow') of seeing a conventional double lift.[50] In forwarding a distinction between initial reactions and deeper experiences, DaOrtiz sought to make something known to the group; namely that we did not properly grasp what to look for in assessing a basic technique familiar to all around the table. His non-verbal pointing accompanying his verbal explanation served to provide a public display for how this lack of understanding could become appreciated. By attending to specific 'domains of scrutiny'[51]

---

50 That is, a sensation generated from the visual dissonance between the identity of the card initially shown in his hand, and the one eventually revealed on the table.
51 Goodwin, Charles. 1994. 'Professional Vision', *American Anthropologist*, 96(3): 606–633.

(the two positions of cards), we were encouraged to appreciate what we visually saw but did not properly register. The demand placed upon us as learners was to search out how we experienced the feeling that two cards were involved. This needed to be done even though all of us had pre-existing knowledge that a double lift involves two cards. Such examples, it could be said, brought home to us how our attention—and therefore the attention of the audiences for our magic—is 'sometimes voluntarily controlled and performed intentionally while it is not always so controlled and not always performed intentionally'.[52] Such were the complicated dynamics presented as at play in appreciating the presence of and sources for naturalness.

A task that we, as learners, faced *after* the masterclass was how to incorporate the teaching into our own practice. As suggested in the description given above, DaOrtiz offers a holistic presentational style; the subtle timing of gestures, the nuances of speech, the forms of the sleights undertaken, the details of bodily comportment, the precise direction of gaze, and many other details besides are purposefully crafted to project a sense to the audience that the behavior is natural. This integrated coordination poses a basic tension: only bringing into discrete, individual elements of DaOrtiz's approach would reduce the prospects for recreating a sense of magic generated through his style. Yet seeking to reproduce his style as a whole takes the risk of sliding into mere copying.

Copying is a noteworthy option because avoiding imitation is a frequent refrain in instructions for beginners. Since each individual has a different personality, many magicians argue that attempts to duplicate another person's presentational style are bound to end in disappointment.[53] High-profile professional magicians have spoken out against how some in the magic fraternity lazily copy well-known artists.[54] In contrast to such typical guidance, DaOrtiz maintains that copying others for their precise timing and movements is a predictable

---

52 Watzl, Sebastian. 2017. *Structuring Mind: The Nature of Attention and How It Shapes Consciousness*. Oxford: Oxford Scholarship Online. https://doi.org/10.1093/acprof:oso/9780199658428.001.0001
53 Nelms, Henning. [1969] 2000. *Magic and Showmanship*. Mineola, NY: Dover: 54.
54 For one discussion of this, read Regal, David. 2019. *Interpreting Magic*. Blue Bike Productions: 209–218. See, as well, Greenbaum, Harrison. *The Insider*. 18 November 2019. https://www.vanishingincmagic.com/blog/the-insider-harrison-greenbaum

first step in being able to work under the influence of others. Indeed, this was the course he adopted in his own development.[55]

Conclusion

As has been said, within the contrived circumstances of modern magic, 'naturalness doesn't come naturally'.[56] Conjurors expend a great deal of effort trying to convince audiences that they are acting in ways that are ordinary, even as others likely suspect that the naturalness is purposefully designed to dissimulate.

With my focus on examining how magic is learnt, in this chapter I have sought to consider how appreciating the quality of naturalness is varyingly presented as possible. At the heart of this has been a basic tension with instructions. If perceiving (un)naturalness came effortlessly, there would be nothing worth pointing out as part of the instructions, since everything worthy of note would be obvious. If perceiving (un)naturalness was not possible at all, there could be nothing to point out.

Instead of either of these extremes, the instructional examples surveyed in this chapter sought to advance skilful bases for perceiving and displaying naturalness to generate feelings and sensations of wonder. Drawing distinctions between looking and feeling, as well as coaching how to move between the two, for instance, served as a basis for grasping what was natural.

In their degree of attention to affect as part of shoring up a sense of naturalness, those surveyed in this chapter would take some issue with psychologist Margaret Wetherell's contention that:

> Because we engage in affective practice all the time, every member of society possesses a wide-ranging, inarticulate, utilitarian knowledge about affective performance: how to enact it, how to categorise it, and how to assign moral and social significance to affective displays.[57]

---

55 DaOrtiz, Dani. 2017. Penguin Dani DaOrtiz LIVE ACT. https://www.penguinmagic.com/p/11142

56 Ortiz, Darwin. 1994. *Strong Magic*. Washington, DC: Kaufman & Co.: 316. For a wider discussion on the need for justified movements in the case of acting, see Lecoq, Jacques. 2000. *The Moving Body*. London: Bloomsbury.

57 Wetherell, Margaret. 2012. *Affect and Emotion: A New Social Science Understanding*. London: Sage: Chapter 4. https://doi.org/10.4135/9781446250945.

At times, the practitioners examined in this chapter would contend that everyone possesses the ability to know how to enact affect, categorize it and assign it significance. These abilities could thereby be labelled as 'ubiquitous' forms of expertise,[58] like speaking your native language. However, the practitioners in this chapter would not always do so. Learning magic—and, in particular, learning modern magic—is also done through consciously questioning our affective understanding.

The next chapter extends this one by considering in more detail how the prospects for seeing what is taking place before one's eyes is varyingly portrayed in conjuring tuition.

---

58 Collins, H. and Evans, R. 2002. *Rethinking Expertise*. Chicago, IL: University of Chicago Press. https://doi.org/10.7208/chicago/9780226113623.001.0001.

5. Proficiency and Inability

Performing Deception opened with a characterization of entertainment magic as *deft contrariwise performance*. The three previous chapters have explored the possibilities, troubles and hauntings associated with the play of opposites in learning conjuring: both becoming closer to and more distant from an appreciation of self; both developing a connection with and recognizing a disconnection from others; both seeking to engineer control of the audience and depending on their lively cooperation; and both cultivating naturalness and pursuing affectation.

This chapter turns to address the interplay of proficiency and inability in practicing and performing magic. Not least because of the reliance on secreted methods, what counts as proficiency in conjuring can be a topic of disagreement. The previous chapter ended by touching on one such matter: is the meticulous imitation of idols by novices a necessary stage of artistic development or a stifling dead-end? Similar questions implicating the place of skill abound. Is the mastering of sleight-of-hand techniques a requisite competency of magicians or not? To what extent can magicians rely on so-called self-working tricks (of the kind set out in Chapter 2)?

Many of those theorizing about magic have argued against the importance of the technical sophistication of tricks.[1] Since what is sought is the ability to elicit feelings of mystery or awe, whether or not artists use demanding sleight-of-hand techniques is neither here nor there. This way of thinking can apply even when magicians evaluate each other. For instance, in February 2021 I entered my first magic competition. Whilst I spent considerable time honing a series of sleights applied to a single standard deck of cards, the winning performance

---

[1] For classic statements along these lines, see Fitzkee, Dariel. 1945. *Magic by Misdirection*. Provo, UT: Magic Book Productions; and Devant, D. and Maskelyne, N. 1912. *Our Magic*. London: George Routledge & Sons.

relied on several specially designed, pre-arranged card decks used one after the other.

At times, however, the ability to perform sleight of hand does matter. For instance, in 2021 I gained membership into the Magic Circle, an international society of professional and amateur magicians. For this, I needed to pass a performance examination. While the routine could include self-working tricks, applicants to the Circle were advised that 'an act consisting entirely of a succession of standard self-working dealer tricks is unlikely to earn you sufficient marks' to pass.[2]

Questions of skill impact directly on questions of identity. Since, as Derren Brown argued, 'any child who can search endlessly for your card in a special deck from a toyshop can call [themselves] a magician',[3] leading figures in the field often vocally question what distinguishes proper conjurors from pretenders. Do conjurors need to develop their effects and presentations to be considered legitimate artists? If they do not, are they no more than band cover artists or, worse, karaoke singers?[4]

Of the many potential areas in which attempts to define proficiency play out, this chapter attends to the manner in which perception underlies claims to proficiency. Perception here refers to how sensory input is identified, interpreted, experienced and, thereby, informs our beliefs about the world. In general terms, magic has an unsettled relationship with the senses. When audiences witness a magician's assistant getting locked into a cabinet and then its doors are opened to reveal emptiness, a contradictory invitation is extended. Audiences are both invited to rely on their sight in a matter-of-fact way and yet also issued with a caution against doing so. As is the case for audiences, so too for newcomers. Learning magic entails honing something of an 'eye' for detail. Becoming proficient with cards, for instance, requires attending to subtleties of their touch, positioning and other qualities. And yet, becoming proficient also entails minding the fallibility of the senses.

How individuals are invited to closely attend to—and come into doubt about—what they perceive is the recurring theme in this

---

2   See Magic Circle. 2017. *Guide to Examinations* (November). London: Magic Circle. See https://themagiccircle.co.uk/images/The-Magic-Circle-guide-to-examinations.pdf
3   Brown, Derren. 2006. *Tricks of the Mind*. London: Channel 4 Books: 34.
4   The former claim being one advanced in Greenbaum, Harrison. *The Insider*. 18 November 2019. https://www.vanishingincmagic.com/blog/the-insider-harrison-greenbaum.

chapter. To do so, I explore questions such as: who can perceive magic performances, properly? How are the skills associated with perception socially distributed? How do experts demonstrate to learners the limits of their perception? The basic orientation for addressing these questions is to treat perception as practical accomplishment involving a host of considerations far beyond our physiology.

Maxims for Magic

As initially outlined in Chapter 2, the importance of experiencing magic from the audience's point of view is a frequent refrain among conjurors. The failure to do so often serves as a basis for professional reprimand. As Darwin Ortiz counselled fellow conjurors in his widely acclaimed book *Strong Magic*, seeing an effect from the audience's point of view 'is something you must always strive for, yet which most magicians fail to do.'[5]

More than just this though, at times he argued that seasoned performers are decisively *worse* than uninitiated audiences in bringing a discerning eye to bear. As Ortiz contended, the 'moral here as elsewhere is that magicians generally are *less perceptive* audiences than laypeople and an unreliable guide as to what constitutes strong magic.'[6] In arguing that conjurors miss what is strong (and fail to recognize that they miss what is strong), he is hardly alone.[7]

With particular reference to Ortiz's *Strong Magic*, this section outlines contrasting claims made about who is capable of perceiving and feeling what.

To begin, an assumption often operating within conjuring discussions is that with greater experience comes a greater discernment.[8] In this spirit, the front dust jacket of Ortiz's *Strong Magic* outlines his lengthy

---

5   Ortiz, Darwin. 1994. *Strong Magic*. Washington, DC: Kaufman & Co.: 76.
6   *Ibid*.: 244. Emphasis in original.
7   For a discussion of these points, listen to Shezam. 2019, October 14. *Erik Tait on Publishing*. Magic Podcast 40. https://shezampod.com/podcast/40-erik-tait-on-publishing-magic/ as well as The Jerk. 2016. 'The Importance of Combining Methods'. http://www.thejerx.com/blog/2016/6/30/the-importance-of-combining-methods
8   As in, for instance, Maskelyne, Neil and Devant, David. 1911. *Our Magic*. London: George Routlege and Sons: Preface. For a theorization of magic along these lines, see Goto-Jones, Chris. 2016. *Conjuring Asia: Magic, Orientalism, and the Making of the Modern World*. Cambridge: Cambridge University Press.

experience as a close-up entertainer and a consultant on crooked gambling methods. His extensive experience, even compared to other professional magicians, is repeatedly evoked as underpinning his authority to justify a 'meta-expertise'[9] in being able to judge other practitioners.

And yet, as indicated above, Ortiz offered several ways in which familiarity with magic can result in a kind of learnt incompetency. Consider some of these ways in more detail. As he argues, instead of prior familiarization with an individual effect resulting in a more refined eye, repetition can result in overexposure that means magicians 'become unable to appreciate just how strong the basic effect really is'.[10] Instead of knowledge of conjuring techniques resulting in magicians being more difficult to fool, the 'knowledge of magic serves only to ossify their thinking'.[11] Relatedly, without preconceived notions about how magic is done and 'because they're not overly concerned with the exact details of methodology, laypeople can more easily see the big picture, and often instinctively go directly to the correct solution'.[12]

My experience chimes with these concerns about learnt incompetency. One paradoxical outcome of practice was in how my development as a learner took me *away* from being able to appreciate the perspective of (lay) audiences. As recounted in Chapter 2, in undertaking my first trick, my situation was in line with that of a naïve spectator: I had no understanding of how the outcome previewed in the instructions would be possible. As I became conversant with the methods for tricks, though, I had to attempt to dissociate what I knew as a learner from what I would experience as a spectator. The result was a bind. The more I learnt, the more grounds I had for doubting the appropriateness of using my mind as an analogic basis for gauging the experiences of others.[13]

For Ortiz, the inability of magicians to judge what counts as strong magic is tied to a second form of learned incompetency. Instead of their extensive experience enabling magicians to judge what works,

---

9   Collins, H. and Evans, R. 2002. *Rethinking Expertise*. Chicago, IL: University of Chicago Press. https://doi.org/10.7208/chicago/9780226113623.001.0001
10 Ortiz, Darwin. 1994. *Strong Magic*. Washington, DC: Kaufman & Co.: 224. See also Earl, Benjamin. 2018. *Roleplayer*. Sacramento, CA: Benjamin Earl & Vanishing Inc.
11 Ortiz, Darwin. 1994. *Strong Magic*. Washington, DC: Kaufman & Co.: 405.
12 Ortiz, Darwin. 1994. *Strong Magic*. Washington, DC: Kaufman & Co.: 405–406.
13 For a related discussion of the need and difficulty of re-appraising tricks, see Kestenbaum, David. 2017, June 30. 'The Magic Show — Act Two', *The American Life*. https://www.thisamericanlife.org/619/the-magic-show/act-two-31

Ortiz claims that experience reinforces a sense of what works best for magicians according to their particular style preferences. The recurring failure of conjurors to recognize this means they can offer poor counsel to colleagues.[14]

Similarly, contrasting arguments are put forward in *Strong Magic* regarding magicians' abilities to see from the audience's point of view. On the one hand, this is presented as a fairly straightforward task, since the audience can be led in various ways by a competent performer. Ortiz offers the instruction, which most readers of *Strong Magic* would likely already be familiar with, regarding how to direct the audience's attention: treat as important what you want the audience to treat as important, and disregard what you want the audience to disregard.[15]

On the other hand, seeing as another is said to be fraught. The kernel of the problem is that magicians scrutinize effects for how they *are* done. What they should instead do, is scrutinize effects for how the audience *guess* they are done.[16] To use an old term in the philosophy of aesthetics, the danger for magicians is that they become 'over-distanced' from their art; preoccupation with technique and stagecraft results in magicians tricking themselves into a form of inattention about what lay audiences perceive.[17] As Ortiz argues, the fascination with handling techniques means that conjurors are insensitive to what matters for lay audiences. Magicians can dismiss the power of effects that are not based on elaborate trickery (for instance, bending spoons with 'the mind'), and actions that magicians know are irrelevant to the performance of tricks (for instance, making sure one's sleeves are rolled up). In contrast, it is the ability to assume the perspective of a naïve spectator that marks Ortiz's expertise.

Furthermore, because of their knowledge and preoccupation with the secreted methods at play, for Ortiz conjurors can spend 'a great deal of time and effort to prove something that isn't even in contention in the audience's mind'.[18] Interjecting explanations and patter where none are needed undercuts the affective power of effects.[19]

---

14 Ortiz, Darwin. 1994. *Strong Magic*. Washington, DC: Kaufman & Co.: 343.
15 *Ibid.*: 37.
16 *Ibid.*: 73.
17 Dawson, Sheila. 1961. '"Distancing" as an Aesthetic Principle', *Australasian Journal of Philosophy* (Vol. 39): 155–174.
18 Ortiz, Darwin. 1994. *Strong Magic*. Washington, DC: Kaufman & Co.: 96.
19 Similarly, some magicians have argued that others too often attempt using advanced card control sleights which require considerable skill, when simpler ones would be

Another area in which experience hampers the prospects for conjurors to gauge the affective power of performances relates to audience reaction. On the one hand, Ortiz looks to the audience reaction as a gauge for what works.[20] Readers are encouraged to review how audiences respond and to search out why they do so. Experience thereby buttresses expertise. And yet, Ortiz also recognizes that some spectators will be too polite to air critical thoughts, and instead engage in a form of counter-deception.[21] Equally, magicians are likely to be too self-absorbed to gauge accurately how audiences are actually reacting.[22]

In contending that experience leads to learnt incompetence,[23] the claims made above are not unique to magic. With time, teachers can lose sight of what is required to learn something new. With time, doctors can become desensitized to what it means to receive a serious diagnosis. With time, politicians can become divorced from the public they intended to serve. And so on. Such claims rely on a form of ironic contrast: what appears to be the case to professionals is really otherwise.[24] Students are not inspired; patients are not at ease; and voters are not stirred. As suggested by the survey in the section, for Ortiz, magic is a thoroughly ironic activity: by their very efforts to become more skillful, conjurors lose the apprehension of their audiences, their peers and themselves. The trick Ortiz needs to pull off given this irony is how to be a taken as authoritative, given the binds he himself sets out.

Another interesting feature of Ortiz's argument and other commentaries on the relationship between experience and expertise is which arguments are *not* made. Beginners—with one foot in the lay

---

more effective due to appearing more natural and expected to audiences. As in Earl, Ben. 2020. *Deep Magic Seminar*. 16 July.
20 Ortiz, Darwin. 1994. *Strong Magic*. Washington, DC: Kaufman & Co.: 342.
21 *Ibid.*: 422. See, as well, Jon Armstrong. *Insider*. https://www.vanishingincmagic.com/insider-magic-podcast/
22 Ortiz, Darwin. 1994. *Strong Magic*. Washington, DC: Kaufman & Co.: 344–345.
23 Magicians have identified other forms of learnt incompetence. For instance, while developing tricks can require high levels of creativity, the willingness of some magicians to latch onto an initial working solution has been said to mean they can be uncreative. Pritchard, Matt. 2021, September 24. *Comments at SOMA Magic & Creativity Webinar*. https://scienceofmagicassoc.org/blog/2021/8/23/magic-creativity-webinar.
24 Schneider, Tanja and Woolgar, S. 2012. 'Technologies of Ironic Revelation', *Consumption Markets & Culture*, 15(2):169–189. https://doi.org/10.1080/10253866.2012.654959.

audience group and one foot in the inner world of conjuring—might be regarded as being in an ideal position for judging what counts as strong magic, what works and so on. However, such a tack has not been evident in my apprenticeship. I cannot recall a single instance in a magic convention, magazine article, online discussion group, 'how to' manual and so on in which beginners have been placed in an elevated or even potentially advantageous position for scrutinizing conjuring routines. Instead, it is only those at the extremes that are presented as able to really judge what is what: lay audiences and topflight professionals.

In the ways indicated in the last several paragraphs, *Strong Magic* offers seemingly contrasting, even directly opposing, claims regarding whether familiarization and experience aids or hinders discernment. This could be taken as presenting an incoherent message that is therefore problematic. However, the presence of opposing ways of thinking is arguably a pervasive feature of everyday and professional advice-giving. For instance, everyday common-sense maxims both suggest:

- 'You're never too old to learn' and 'You can't teach an old dog new tricks';
- 'Wise people think alike' and 'Fools seldom differ';
- 'Hold fast to the words of your elders' and 'Wise individuals make proverbs; fools repeat them'.

As psychologist Michael Billig and colleagues have argued, the existence of contrasting ways of approaching questions about how to act is widespread. More than this though, it is unavoidable.[25] It is unavoidable because the availability of opposing ways of thinking provides the very basis for individuals and groups to think through what should be done in a specific situation. In the case of assessing magic, for instance, the extensive experience of a conjuror might well justify confidence about how well they can scrutinize routines. However, such experience might well justify caution in a different case. For some effects, the adage 'If it ain't broke, don't fix it' might be deemed to apply. At other times, the imperative for innovation might hold sway.

---

25 Billig, M. 1996. *Arguing and Thinking*. Cambridge: Cambridge University Press; and Billig, M., S. Condo, D. Edwards, M. Gane, D. Middleton and A. Radley. 1989. *Ideological Dilemmas*. London: Sage.

At times, Ortiz himself notes the scope for his guidance to be countered. In offering evaluations of card effects, he also contends, 'I hasten to add that I know full well that for every statement I've made there is at least one really great card effect that contradicts it'.[26] Instead of his advice being fit for all, he goes on to say: 'However, the prejudices I've described above are right for *me*. Following these biases has helped give my performances a distinctive and consistent look...'.[27] With such qualifications basic on aesthetic judgements, the overall evaluation given is presented as stemming from a particular way of thinking about magic, one that readers might be wise to heed if they are aligned with Ortiz's style preferences...but one they might choose to ignore, too.

Science of Magic

Ortiz's ability to assess the perceptiveness and reliability of other magicians derives from his real-world experience and professional achievements. In this way, he assumes the status as a kind of connoisseur. His intensive and attentive immersion into card magic has enabled him to appreciate aspects of magic that pass other professionals by and to skirt around trap doors into which others keep falling.

Whilst practical experience has traditionally served as the chief grounding for claims to expertise in entertainment magic, it is not the only one. For many decades, fields of science have sought to explain why sleights and other forms of trickery prove so hard to detect.[28] In recent years, under the label 'The Science of Magic,'[29] renewed interest has emerged in utilizing magic effects as experimental stimuli in efforts to characterize visual perception and cognitive heuristics.[30] One review summed up the principles identified through this latest phase of research as:

---

26 Ortiz, Darwin. 1994. *Strong Magic*. Washington, DC: Kaufman & Co.: 308.
27 *Ibid*. Emphasis in original.
28 Lamont, P. 2006. 'Magician as Conjuror', *Early Popular Visual Culture*, 4(1): 21–33.
29 https://scienceofmagicassoc.org/home
30 Kuhn, G. 2019. *Experiencing the Impossible*. Cambridge, MA: MIT Press. ttps://doi.org/10.7551/mitpress/11227.001.0001; as well as Kuhn, G., Caffaratti, H., Teszka, R. and Rensink, R.A. 2016. 'A Psychologically-Based Taxonomy of Misdirection'. In: *The Psychology of Magic and the Magic of Psychology* (November), Raz, A., Olson, J. A. and Kuhn, G. (Eds). https://doi.org/10.3389/fpsyg.2014.01392.

First, some things, though directly in a person's line of sight, are not perceptible at all. Second, people do not consciously perceive everything that can be perceived. Third, what is consciously perceived depends upon attention. Individuals will fail to see even what is in their direct line of sight or fail to feel an easily perceptible touch if their attention is elsewhere. Fourth, people sometimes misinterpret what they perceive. Fifth, individuals' memories fail in ways that permit changes to occur before their eyes that they do not consciously perceive. Sixth, these failures can be regularly and lawfully produced by specific manipulations of individuals' perceptual and sensory systems.[31]

In short, what is observable depends on the means of observing. As a result, how we believe we observe is often not how we observe in practice.

A further goal in The Science of Magic is to take the counterintuitive lessons learned to improve how magic is performed. One reason this is possible is because—despite being adept at harnessing perceptive and cognitive limitations—conjurors are often as susceptible to being fooled as anyone else. This is so, not least, because magic effects can rely on automatic visual and cognitive processes that are not directly noticeable.[32]

In *Experiencing the Impossible*, psychologist and magician Gustav Kuhn marshalled findings from The Science of Magic to propose how research could advance performances. Herein, even while magicians know how to exploit perceptual failures, he argued 'I do not think they fully appreciate their magnitude, nor do they fully understand why these changes occur.'[33] Take the example of change blindness—the perceptual phenomenon whereby modifications can be introduced in visual stimuli without observers noticing. As Kuhn argued, many professionals can be surprised by the scope for change blindness. As a result:

Inasmuch as all of us (including magicians) intuitively overestimate the amount that we consciously perceive, magicians could be developing

---

[31] Villalobos, J.G., Ogundimu, O.O., and Davis, D. 2014. 'Magic Tricks'. In: *Encyclopedia of Deception*, T. R. Levine (Ed.). Thousand Oaks: Sage: 637. https://doi.org/10.4135/9781483306902.

[32] Ekroll, Vebjørn Bilge Sayim, and Wagemans, Johan. 2017. 'The Other Side of Magic', *Perspectives on Psychological Science*, 12(1): 91–106. https://doi.org/10.1177/1745691616654676.

[33] Kuhn, G. 2019. *Experiencing the Impossible*. Cambridge, MA: MIT Press: 220. https://doi.org/10.7551/mitpress/11227.001.0001.

bolder and more daring techniques. Magicians typically assume that attention simply refers to where you look, but our work shows that people often miss seeing things that are right in front of their eyes.[34]

For instance, not only can spectators miss it when the color of playing cards is changed, because they are drawn to look at a magician's face, they are just as likely to miss the color change when looking at the cards.[35] The extent to which people can miss what is taking place in front of them means that even psychologists such as Kuhn are surprised by the scope of what can go undetected.

In The Science of Magic then, research data (rather than experience or status) is advanced as the ultimate gauge of perception. This is so because no one—lay spectator, veteran magician or experimental psychologist—can fully appreciate from their everyday experiences the fallibility of our senses.[36]

Accounting for Perception, Building Proficiency

In terms of my development, reading professional magicians like Ortiz, as well as research scientists like Kuhn, provided concepts and theories for interpreting my observations of conjuring and undertaking experimentations as part of shows.

In the remainder of this chapter, I want to engage with the themes of proficiency and perception in both professional and research literature, but with a particular starting concern. Following in the tradition of ethnomethodology-related analysis of sight by Michael Lynch, Charles Goodwin, Tia DeNora, and others,[37] my concern is with how

---

34 *Ibid.*: 221. https://doi.org/10.7551/mitpress/11227.001.0001.
35 A finding which Kuhn and others elaborate in Kuhn, Gustav, Teszka, Robert, Tenaw, Natalia and Kingstone, Alan. 2016. 'Don't Be Fooled! Attentional Responses to Social Cues in a Face-to-Face and Video Magic Trick Reveals Greater Top-Down Control for Overt than Covert Attention', *Cognition*, 146: 136–142. https://doi.org/10.1016/j.cognition.2015.08.005.
36 A conclusion that, while radical, is also in line with many past commonplace orientations to sight; see Clark, Stuart. 2007. *Vanities of the Eye*. Oxford: Oxford University Press. https://doi.org/10.2752/175183409x12550007730345.
37 Lynch, Michael. 2013. 'Seeing Fish'. In: *Ethnomethodology at Play*, P. Tolmie and M. Rouncefield (Eds). London: Routledge: 89–104; Goodwin, C. 1994. 'Professional Vision', *American Anthropologist*, 96(3): 606–633; DeNora, Tia. 2014. *Making Sense of Reality*. London: Sage. https://doi.org/10.4135/9781446288320; and Coulter, J. and Parsons, E.D. 1991. 'The Praxiology of Perception', *Inquiry*, 33: 251–272.

determinations about who can perceive what are advanced as part of specific interactions. In Chapter 4, for instance, I examined how conjuring instructors marshalled distinctions between 'looking' and 'feeling' as a way of sensitizing learners about how to appear natural. This chapter extends that analysis by asking when and how the ability to perceive is made relevant within specific settings. In particular, I examine two types of interactional activities—performances and face-to-face instruction—for how the spoken word, gestures, gaze and other actions organize the place of perception and the abilities of those present.

Let me begin through a personal example.

In terms of the performances, Chapter 3 discussed the small group sessions I started in 2018. In total, 30 sessions were recorded. Particularly early on into running these events, I had little experience in conjuring. Although the self-working tricks in the first 13 sessions did not require sophisticated card sleights, one of the nine did require pushing a card out of the deck to glimpse it, and another entailed covertly turning over a deck. Almost all of the nine effects in the second set of ten sessions involved one or more physical sleights—false shuffles, lifting multiple cards, forcing participants to select a predetermined card, etc. On some occasions, too, when the cards were out of the required order, I needed to rearrange them at the table without arousing suspicion.

In their own way, these recorded sessions realized the call by Kuhn to devise bold occasions for testing perception. This was so not because of the technical sophistication of the methods for the effects, but because of my lack of experience. Due to my lack of abilities, I expected that the jiggery-pokery with the cards would be frequently detectable.

Repeatedly in our discussions, participants offered unprompted explanations for how the effects were accomplished. In addition, I deliberately asked them for their thoughts. And yet, rarely did participants forward (even partially) accurate identifications. While what counts as verbally recognizing a relevant element of the methods for a trick is open to interpretation,[38] I would put the number of such occasions across the first 23 sessions (so the initial three routines) somewhere in the high single digits. These experiences are in line with the overall claims made about perception and cognition within The Science of Magic.

---

38   For one breakdown of forms of explanation, see Smith, W. et al. (forthcoming). 'Explaining the Unexplainable: People's Response to Magical Technologies'.

Across all 30 sessions, another absence was of note. In only two sessions were suggestions voiced by participants that their perceptions could be significantly fallible. Neither were more general claims offered that what was observable significantly depended on the means of observing or reporting. Instead, participants made much more delimited claims, such as that sight can be misdirected (for instance, see Excerpt 3.4, Line 22).[39]

In brief, participants accounted for the unfolding scene through a realist language according to which a familiar world is out there, independent of our actions, and delivered to our consciousness (as we attend to it via our senses).

Through such accounting, the scene was rendered what Melvin Pollner called 'mundane'. In his classic study, Pollner identified 'mundane reasoning' as a ubiquitous form of constructing the world wherein individuals 'experience and describe themselves as 'reacting to' or 'reflecting' an essentially objective domain or world'.[40] Within the traffic court proceedings he examined, for instance, witnesses to an incident could offer radically divergent accounts of what took place. Judges seeking to adjudicate 'what happened' were thus in a position of striving to determine the facts while also being reliant on conflicting observations.[41] Pollner detailed how courtroom judges could both determine the 'facts of the matter' based on divergent accounts, as well as preserve the starting presumption that there was an essentially out-there, ordinary and objective world to be found that could be taken to exist independently of anyone.

Participants in my sessions accounted for the unfolding scene through similar realist conventions. Instead of calling into question the determinacy of perception, participants proposed stock explanations for how effects were accomplished. These echoed popular understandings of the methods of magic and were overwhelmingly incorrect or, at best, referring to highly general principles. Erroneous explanations included:

---

39   In this case, the contention was made that it is possible to fail to see what is in one's direct line of sight.
40   Pollner, M. 1987. *Mundane Reason*. Cambridge: Cambridge University Press: xv.
41   For a historical analysis of how conflicting observations have been alternatively dealt within judicial settings, see Saltzman, Benjamin A. 2019. *Bonds of Secrecy*. Philadelphia, Pennsylvania: University of Pennsylvania Press. https://doi.org/10.9783/9780812296846.

psychological priming, the placement of cards up sleeves, the use of hidden mirrors and (in the case of the self-working tricks) physical sleight dexterity.[42]

An important facet of this attribution is what it meant for our identities. Relating to the self-working tricks, the belief that I was engaging in (and even that I could engage at all in) covert sleight-of-hand movements invested in me technical abilities I did not possess (for instance, see Excerpt 3.4, Lines 16, 20). More generally, in the absence of accounting for magic through reference to our shared limited human perceptual and cognitive capabilities, attempts to explain what was taking place repeatedly evoked *my* skills, *my* plans, *my* doings, and so on.[43] When I was rendered as possessing extraordinary skills, the ordinary status of the world did not come under scrutiny. Almost without exception, across the 30 recorded sessions, participants did *not* voice any concerns about the fallibility of perception and cognition.

Two of the 30 recorded sessions proved to be the exceptions. Within these sessions, reference to the limits of perception related to 'perceptual' or 'inattentional' blindness. This blindness refers to the way an object in plain sight can be rendered hidden because attention is focused on other objects in our field of vision. In both sessions, the iconic example of the 'invisible gorilla' psychological experiment was cited by participants.[44] While this experiment was only mentioned in passing during one session, in the other it figured as a recurring reference point. This latter audience consisted of three philosophers of mind, all versed in the science of human perception. One trick entailed a participant signing a selected card. Later, that card was selected again, and this time I signed it as well and then returned it to the deck. Several minutes later, the card

---

42 As such, participants' prior familiarization with magic in general served to bolster specious interpretations of what was taking place in a specific instance. In this way, with more familiarity with the methods in magic came scope for participants to entangle themselves with their own explanations.

43 To distinguish these interactions from other historical periods, no claims were made about the illusionary qualities of nature nor the possibility that demonic forces were manipulating perception of the kind discussed in Clark, Stuart. 1997. *Thinking with Demons*. Oxford: Oxford University Press; and Clark, Stuart. 2007. *Vanities of the Eye*. Oxford: Oxford University Press.

44 As recounted in Simons, Daniel J. and Chabris, Christopher F. 1999. 'Gorillas in Our Midst: Sustained Inattentional Blindness for Dynamic Events', *Perception*, 28: 1059–1074. If you are reading this because you don't know the gorilla experiment, then you must visit: http://www.theinvisiblegorilla.com/gorilla_experiment.html

signed by both of us appeared inside a capped water bottle on the table. The following exchange ensued after P2 discussed recently rewatching a version of the gorilla experiment:

Excerpt 5.1—Session 3

No	Direct transcript	Non-verbal actions
1	P1: But that kind of a trick, if you focus on that maybe you are a little, but with this kind of thing it makes me feel, oh crazy, because, it, there is a lot of time you have to, it takes	
2	P1: [to]	P1: gestures toward bottle; then makes opening bottle gestures.
3	P3: [Yeah]	
4	P1: do this, to open.	
5	P3: Yeah, yeah	
6	P1: So you have to	
7	P3: Yeah, yeah.	
8	((multiple voices talking over each other))	
9	P2: That's how inattentive we were. That's how inattentive we were.	
10	((laughter, multiple voices))	
11	P3: I mean it is good that he pulls the card and then signs.	
12	P1: Maybe it was since, it	
13	P1: [was there since]	
14	P3: [NO, NO, I don't know.]	
15	P1: When we started	
16	P2: But he could have easily taken the bottle down like from the side.	P2: Gestures with right hand moving down over the edge of the table
17	P1: No, NO, NO	

5. Proficiency and Inability   113

No	Direct transcript	Non-verbal actions
18	BR: I did it right here in the middle of the table. Was this your card?	*Brian: Energetic simulation of twisting a bottle cap open at the center of the table*
19	P2: REALLY. REALLY. If you like played the tape and that is what happened I won't, I would not be surprised.	
20	((Group laughter))	
21	BR: I push the card down.	*Brian: Simulates pushing card into a bottle at the center of the table*
22	P2: I was, I was <u>so</u> inattentive. I was like so into like shuffling	
23	P2: ((laughter))	
24	P2: You could have put on a gorilla costume.	
25	P3: Yeah, yeah, yeah. Who would have noticed? ((laughter)) He is naked.	
26	((Group laughter))	

In this unfolding interaction, a sense of what happened was reconstructed. Inattentional blindness became an explanation that not only provided a sense of how the card-in-the-bottle feat was accomplished, but also a sense of participants' flawed perceptual capabilities, as well as my practiced abilities.

However, interestingly, this effect did *not* rely on inattentional blindness. I should say it did not in any significant sense. The card-in-the-bottle was only readily visible on the table to the participants for several seconds before I directed their attention to it. Even if they had seen it at the start of this period, the intended goal would have been achieved.[45]

---

45 Funnily enough, the 'Card in Bottle' instructions I'd learnt *had* suggested making use of inattentional blindness by prescribing the card-in-the-bottle be placed in view for a lengthy period. I did not take this path, though. As a relative beginner, I

In making the concept of 'inattentional blindness' relevant to our moment-to-moment interactions, these participants thereby created a sense of what was going on and the identities and capabilities of those involved. Through mobilizing their existing knowledge about the psychology of perception, they came to reinforce a sense that they were perceptually flawed and that I, as a performer, skillfully harnessed this incapability. In other words, unlike the other recorded sessions, in this, the notion that the world was *not* 'out there' as a given phenomenon provided that basis for making magic and attributing heightened competencies.

Schooling Perception

Relating to how perceptual limits were made relevant to interactions, the previous section focused on how participants in routines co-performed and inflated my capabilities as a novice. This section turns to a different kind of activity in which perceptual abilities were at stake, namely face-to-face tutorials.[46] I will consider how the limits of perception were made witnessable.

Previous research across diverse fields of art and craft suggests that face-to-face teachings of bodily skills are often characterized by embodied forms of epistemic and charismatic authority in which expertise is shared through gesture, repetition and sensory apprehension.[47] For instance, in the case of operatic masterclasses, teachers engage in varied forms of hands-on instruction so as to demonstrate how students should comport themselves. That can mean gesturing to highlight precise movements necessary to breathe properly. It can mean teachers more or less subtly

---

adopted a far more cautious strategy for getting the card in the bottle. This meant the card in the bottle was only able to be seen by the participants for a short time.

46 Jones, Graham. 2011. *Trade of the Tricks*. London: University of California Press. https://doi.org/10.1525/california/9780520270466.001.0001.

47 E.g., Evans, J., Davis, B. and Rich, E. 2009. 'The Body Made Flesh: Embodied Learning and the Corporeal Device', *British Journal of Sociology of Education*, 30(4): 391–406. https://doi.org/10.1080/01425690902954588; Ivinson, G. 2012. 'The Body and Pedagogy: Beyond Absent, Moving Bodies in Pedagogic Practice', *British Journal of Sociology of Education*, 33(4): 489–506. https://doi.org/10.1080/01425692.2012.662822; Marchand, T.H.J. 2008. 'Muscles, Morals and Mind: Craft Apprenticeship and the Formation of Person', *British Journal of Educational Studies*, 56(3): 245–271. https://doi.org/10.1111/j.1467-8527.2008.00407.x.

re-positioning students' bodies.[48] Or it can mean teachers undertaking and describing actions with their own bodies that students are meant to mimic. In such instructions, showing and telling are intertwined. It is through such acts of showing and telling that teachers affirm their proficiency.

The previous argument set out in *Performing Deception* suggests displaying proficiency in magic is likely to be a tricky endeavor. Learning magic requires utilizing perceptions to discern what is shown and told, but learning magic also involves coming into an appreciation of the limits of our perceptions. How, then, are the senses explained, honed and disregarded as part of face-to-face student-teacher training? How are appeals to perceptions used to evidence, demonstrate and challenge notions of who can appreciate what is taking place? How do teachers establish their authority to speak for others' experiences?

The remainder of this chapter addresses these questions by revisiting the masterclass I attended with renowned magician Dani DaOrtiz (see Chapter 4). I want to draw out how the instructions cultivated sensitivities for moving between varied orientations to our perceptions.

As background comments, the instructional sessions as part of the masterclass largely consisted of us (a group of seven students) sitting around a table physically orientated toward DaOrtiz (see Figure 5). As an instance of masterclass training, this event differed from many others in that we as students were not asked to perform so that DaOrtiz could offer appraisals.[49] Instead, he performed a copious number of effects, worked through the mechanics for many of those effects with us as students, and we listened to and asked questions about the wider psychological theorizing that informed his chaotic style. Through such activities in which Dani held sway over the floor, we were invited to witness his performance skills, the quality of which we gauged individually. As well, the bedazzled reactions of other students, the applauses we mutually created, as well as the collective laughter that abound reinforced a sense of his skills in producing magic. As another

---

48   Atkinson, Paul, Watermeyer, Richard and Delamont, Sara. 2013. 'Expertise, Authority and Embodied Pedagogy: Operatic Masterclasses', *British Journal of Sociology of Education*, 34(4): 487–503. https://doi.org/10.1080/01425692.2012.723868

49   As in Ruhleder, K., and Stoltzfus, F. 2000. 'The Etiquette of the Masterclass', *Mind, Culture and Activity*, 7(3): 186–196. https://doi.org/10.1207/s15327884mca0703_06.

measure of the authority he achieved, we as students rarely verbally queried his contentions.

Fig. 5 — A Chaotic Practice Table[50]

One recurring theme was his invocation of the need to distinguish between the magician's and the spectator's point of view. In line with other practitioners already surveyed in this book, developing an appreciation for the latter was presented as vital. By way of understanding how experiencing magic from the spectator's point of view related to proficiency and perception, below I want to attend to how the masterclass combined notions of:

- what was directly perceptually accessible and what required refined acumen;
- the relevance (or not) of prior familiarity with magic;
- demonstrating and telling.

The masterclass began with a display of competency and charisma. Our first session together consisted of over two hours of DaOrtiz performing seemingly effortless table-based card effects in his characteristic chaotic style. Again and again, such effects led to expressions of bafflement,

---

50   Photo: Brian Rappert (28 July 2019).

statements of 'Wow', and looks of incredulity. As instances of modern conjuring, these effects repeatedly traded on the notion that we as spectators were being shown what we needed to see regarding how the cards were being handled. And yet, through the improbable feats undertaken, it was also made clear, too, that much was hidden

At times, DaOrtiz used repetition to illustrate that our conjuring know-how as students in a masterclass did not prevent us from being fooled. For instance, the masterclass included a variety of 'situational effects' that were meant to function as part of the build-up to major effects. Among those effects included a playing card that repeatedly appeared in a seemingly empty box. The masterclass also included the recurrent use of some sleights. We were repeatedly invited to freely select any card from a deck, but DaOrtiz ensured we selected the one he wanted us to pick by using a technique called 'forcing'. Again and again. Through the process of repetition, we were invited to consider the limits of what we could discern even with our pre-existing knowledge of card sleights in general and our knowledge that DaOrtiz was performing sleights in these instances.

The masterclass also varied in the types of verbal statements that accompanied effects. As instances of modern conjuring, DaOrtiz's performances regularly incorporated patter that acted to purposefully direct attention. For example, he offered statements such as 'You remember you shuffled the deck'. Many such contentions were false and intended to mislead (see below). At other times though, DaOrtiz's statements functioned to highlight what was taking place so that the chaotic happenings could later be (partially) reconstructed from memory. In such instances, instead of us as an audience simply being able to take everything in, we needed assistance from him to properly attend to the scene at hand.

Just as the performances entailed a play between the achievement of public visibility and need for discriminating attention, DaOrtiz's explanations for effects could employ nuanced plays. For instance, a recurring teaching technique he used was to perform an effect and then critique that performance from an imaginary spectator's point of view. In this way, even while the students present were spectators to the magic, we were not regarded as being able to judge the displayed effects properly. Instead, DaOrtiz's teachings pointed us towards what might well not be adequately appreciated. To recount one instance, in the masterclass the power of direct tricks that do not require spectators to process significant amounts of information was underscored. To illustrate what counted as 'too much information', DaOrtiz devised the following display, in which a card inexplicably moves between two piles on the table:

Excerpt 5.2—Masterclass

Direct transcript	Non-verbal actions
If the spectator do two piles and the card appear in this pile, don't divide the focus. Because if I do that.	*Cuts the deck into two piles. Points to one pile. Puts piles together and picks up combined deck.*
	Places deck back on table.
Cut	*Student cuts deck into two piles, right (#1) and left (#2).*
OK	
Ah, take any card.	*DaOrtiz picks up right pile (#1) and spreads it in his hands. Student takes a card.*
Alright	*DaOrtiz moves left pile (#2) further to the left of the table.*
Ah, put the card here	*Splits pile initially on the right (#1) in hands. Student puts card in the middle gap.*
Do you remember your card? OK	*DaOrtiz shuffles pile in hand (#1), places it back on the right side of the table.*
Now, pa, pa, pa, pa	*Turns left to pick up the initial left side pile (#2).*
Now I do here. Tagata, tagata, tagata, tagata, tagata, tagata, tagata, tagata	*Shuffles pile #2*
Ahhhm can you take the packet please.	*Turns back right. Gives pile #2 to student.*

Direct transcript	Non-verbal actions
And now look, I try, try to travel. It is not here.	DaOrtiz Rubs hands together. Gestures above pile #1 on table. Spreads pile #1 face up.
And now one card is, uh, here.	Takes pile #2 from student and spreads cards to identify chosen card in pile #2.
This is a s*#t because one pack is here before there, now here. I don't, what is happening here?	Hand arms open. Points in multiple directions using both hands. Waves with both arms. Open arms.
I don't understand. Look, you like a magician say, wow, look, my transfer, my palm, were unbelievable. The spectator say, understand nothing.	Places hands on chest. Performs hand movements simulating sleights. Right palm opens.

Herein, through his uttered words and visible movements, DaOrtiz sought to perform an effect. We, as students, looked on. More than this, he sought to make visible and felt certain aspects of what was performed that we might not have adequately noticed. In other words, how the actions of the magician can be 's*#t'. He did so by drawing our attention to how spectators' attention can be divided. As he contended, a magician might well not appreciate the problems of the performance because of their preoccupation with the execution of physical techniques. As a student-spectator, I took this display as both inviting us to experience that the trick was flawed but indirectly cautioning us how we—as magicians—might well be oblivious to its faults because of our inability to recognize what is in front of us. At conferences, lectures and in training instructions, I have experienced many such fraught demonstrations that both invite and question attention.

While the instructions above entailed crafting a trick in such a way that we could experience what was being pointed toward, there were many occasions in the masterclass that did not involve any *direct* acts of showing. In addressing how to deal with audiences' unexpected actions, in evoking a sense of the contingencies of live performances, in proposing how we would later recall the effects he performed and in other respects, there was no straightforward way DaOrtiz could demonstrate his claims to us there and then on the table. Instead, we as students were asked to imagine, simulate or otherwise speculate. In doing so, we also took on various roles. This included naïve, discerning and belligerent spectators, as well as the role of skilled and novice magicians. Consider one example. After a query from me about how he was using words to affect the actions of spectators, this exchange followed:

Excerpt 5.3—Masterclass

No	Direct transcript	Non-verbal actions
1	DD: For example. Ay, yeah, yeah. Is difficult when you are not in context, because I need to be in a <u>trick</u>.	
2	BR: Yeah,	
3	BR: [yeah	
4	DD: [But it does not matter. OK. I tell you, ahhhh, we shuffle the deck. OK. Cut and complete.	BR: *Cuts deck and then brings pile on top of each other.*
5	DD: And square. Very good?	BR: *Squares deck.*
6	BR: Yes.	
7	DD: You remember, shuffled the deck. And you cut and complete, right? OK.	BR: *Says nothing.*
8	DD: And what the people listen and the people feel is he shuffled the deck. He cut and complete. But he never shuffled. I shuffled. Why? Because he <u>say</u> yes. But why he say yes? Because he say yes to the last part of my question.	
9	BR: Ah, huh, huh, huh	

5. Proficiency and Inability

No	Direct transcript	Non-verbal actions
10	DD: I, if I say, you shuffled and compete and you shuffled cut and complete, right? He say, <u>no</u> because he feel, the two things, shuffle and cut, is in the same sentence. You shuffled the deck and cut and complete, <u>right</u>? And he say, NO. I shuffle but, I cut but I do not shuffle. But what happen if I say, you shuffled, and now I put exclamation. You shuffled the deck <u>and you cut and complete, right?</u>	*DD turns to face a different student.*
11	BR: Hmm.	
12	DD: He tell me YES but in the last part. Not in the beginning.	

In this passage, DaOrtiz addresses how to get audience members to state and even feel for themselves that a deck has been both cut and shuffled by a spectator. Securing such a conviction is advantageous because it undermines the prospect that the audience will believe that an effect could be the result of the conjuror's dexterity with the cards. Within this description, DaOrtiz calls for a complex set of perspectival movements on our part as students, in which what is perceived is the outcome of our interactions together around the table. To give my interpretation of what he called for:

- In Lines 1 DaOrtiz began by offering meta-commentary that qualified what was about to be displayed. This suggested the actions that followed could not simply be taken on their own but need to be somehow contextualized within the doings of a trick. However, in Line 4 he went on to state that the de-rooted status of what was to follow did not matter, a qualification that placed a further question mark over what was about to be shown.
- Line 7 posed a question to me about whether I remembered the deck had been shuffled and cut. However, no affirmative response was given. Owing to the artificial conditions under

which this question was posed, it seems unlikely that one was expected by DaOrtiz. Rather than focusing on what I or others thought about the manipulations to the deck, Line 8 shift to evoke a sense of what a generic audience would hear and feel. We as students were asked to move from our appreciations of the situation at hand to put ourselves in the place of such generic spectators. As such, the elaboration in Line 8 provides a way of making retrospective sense of the question in Lines 7.

- In Line 8 DaOrtiz carried on under the assumption that an affirmative response was given. Herein it was suggested that audiences will be influenced by the positive response of the questioned spectator. Then the explanation for why the spectator says 'yes' is provided toward the end of Line 8: the spectator is responding to the last part of the two-part question posed in Line 7. At this moment we as students were asked to speculate how this might be the case and why it might matter. Doing so called for us to remember back to the specifics of what was said, even as those specifics were meant to lead us astray.

- In Lines 10, the meaning of the 'last part of my question' became clearer because the previous articulation of the question (Line 7) is described as including both propositions (you shuffled and you cut and complete) in the same breath. As DaOrtiz contended, when taken as spoken together, the truth status of both claims was interpreted as relevant for the spectator. As such, DaOrtiz suggested that a spectator will decline the suggestion that they shuffled because they did not. Line 10 repeated the bundling of the two propositions together. However, at the end of Line 10 the second proposition of cutting the deck was verbally emphasized. Now being clearly drawn to the contrast provided by the emphasis, we as students were asked to recognize how the two propositions would be interpreted differently (even if my response in Line 11 did not offer a clear affirmation to the question posed at the end of Line 10).

In this segment, as elsewhere in the masterclass, we as a group of students and a teacher interacted in ways that sought to provide retrospective meaning to what had already taken place that thereby also conditioned how meaning was meant to be made of subsequent events.

After a further exercise in the power of purposefully sequencing and delivering questions to spectators than what is given in 5.3, DaOrtiz would argue that with such techniques you could do 'anything you want'.[51] Despite what might be taken as the speculative and counterfactual status of the demonstration, the contentions forwarded were as persuasive to me during the masterclass as they remained so in relistening to the recordings many months afterward.

To return to the wider theme of what was made visible in the masterclass, at other times, DaOrtiz simply told us what we would experience without seeking to demonstrate his claims. He compared the aesthetic merits of different ways of lifting cards, he contrasted the affective potential of similar effects, he suggested what cannot be visually perceived in a particular situation, and so on. In doing so, DaOrtiz told us what we needed to appreciate rather than leaving it to us to derive our own conclusions or rather than explicitly seeking confirmation (see, for example, Chapter 4 regarding the 'feel' of a double-lift, pages 93–95).

The previous paragraphs speak to some of the ways perception was positioned in the masterclass. Within our moment-to-moment interactions, a sense of experience as shared and diverse was conveyed through verbal and non-verbal actions. As I have come to understand it, part of the demand of learning magic is to be able to move between varied orientations to sensorial experiences. Those orientations entail recognizing what is readily accessible, appreciating what requires refined judgement, perceiving with foreknowledge, disregarding foreknowledge, watching what is demonstrated, disregarding what is shown and imagining what is not shown. The ability to move between such orientations is a crucial form of competency.

As I have come to understand it, too, part of the complex and contradictory demand of being regarded as an authority figure like

---

51 For an analysis of verbal misdirection in teaching magic, see Jones, Graham and Shweder, Lauren. 2003. 'The Performance of Illusion and Illusionary Performatives: Learning the Language of Theatrical Magic', *Journal of Linguistic Anthropology*, 13(1): 51–70. https://doi.org/10.1525/jlin.2003.13.1.51.

Dani DaOrtiz is to be able to account for what takes place, what does not take place, what is real and what is imaginary.

6. Truth and Deception

Previous chapters examined how magic is learnt through considering various engagements: reading 'how to' instructional books, watching online tutorials, participating in face-to-face tuition and offering small group performances. This chapter turns to another resource that can be formative for novices and adepts alike: autobiographies. Such publications purport to offer aspiring conjurors a peek backstage. More than this, though, they also provide readers with exemplars for what it means to be a magician[1] and a path for how to become one.[2]

As a form of writing, autobiographies are often premised on revealing hidden or little-known truths. Even if the author might be familiar to their readers, the appeal of the genre often rests on disclosing what is surprising, noteworthy, extraordinary and so on. Shared confidences, inner motives and hidden struggles are all commonplace components for life histories. Through doing so, autobiographical insights often rub up against what was hitherto generally understood, or they can offer a view of what was not widely seen—or both. For instance, the journalist Ian Frisch's *Magic Is Dead: My Journey into the World's Most Secretive Society of Magicians* not only provides an insider account into a grouping of elite magicians but also their (and his) little appreciated backstories of adversity.[3]

As forms of self-disclosure, autobiographies typically rest on claims to authenticity. Authenticity, though, is an accomplished quality. An author needs to demonstrate their genuineness, lest readers harbor suspicions that their story is a yarn concocted to garner prestige or to settle old scores.

---

1 Matters both analyzed and enacted in Frisch, Ian. 2019. *Magic Is Dead: My Journey into the World's Most Secretive Society of Magicians*. New York: Dey St.

2 Or not at times, listen to Shezampod. 2020. *Podcast 54—Catie Osborn on Shakespeare and Tips From an Entertainment Director*. https://shezampod.com/series/shezam/

3 Frisch, Ian. 2019. *Magic Is Dead: My Journey into the World's Most Secretive Society of Magicians*. New York: Dey St.

© 2022 Brian Rappert, CC BY-NC 4.0   https://doi.org/10.11647/OBP.0295.06

And yet, in the case of conjuring, much of the fascination and intrigue with magicians centers on their recognized ability to mislead.[4] As conjuring relies on honing forms of guile, dissimulation, deceit, simulation, hoodwinking and the like—even while audiences are expecting guile, dissimulation, deceit, simulation, hoodwinking and so on—attempts by magicians to convince others they are genuinely revealing themselves in autobiographies are built on somewhat shaky foundations.[5]

In this chapter, I offer a reading of the autobiographies of leading figures in entertainment magic that takes the management between authenticity and phoniness as its starting point. In doing so, I consider a few questions that inform a sense of what it means to have skill as a magician: what importance do authors invest in their accounts being truthful? How do they fashion autobiographies such that they can hold together evidence of their genuineness with evidence of their ability to mislead? How do conjurors advance notions of right and wrong, even as they recount how they deceive?

Cave Historian

As a project of learning, my pathway into the life stories of magicians did not begin with reading autobiographies, but instead reading historical studies of entertainment magic. In the history of Western magic, doubt that conjurors might—just might—have been less than fully earnest in writing about themselves goes back a long time.[6] In his wonderfully rich book *Performing Dark Arts: A Cultural History of Conjuring*, Michael Mangan spoke to the trepidations that should be associated with relying on the tales of conjurors when he concluded: 'The most realistic way to think about magicians' own accounts of their lives, careers and tricks is to consider them as extensions of their stage acts—as a particular kind of "performative writing"'.[7]

---

4 Avner Insider: https://www.vanishingincmagic.com/insider-magic-podcast/
5 See Allen, Jonathan. 2007. 'Deceptionists at War', *Cabinet* (Summer). http://www.cabinetmagazine.org/issues/26/allen.php
6 See Steinmeyer, Jim. 2003. *Hiding the Elephant: How Magicians Invented the Impossible and Learned How to Disappear*. New York: Carroll and Graf.
7 Mangan, Michael. 2007. *Performing Dark Arts: A Cultural History of Conjuring*. Bristol: Intellect.

Perhaps one of the most prominent such instances relates to the 19th century French conjuror Jean-Eugène Robert-Houdin. Sometimes referred to as the 'King of Conjurors', Robert-Houdin has come to be regarded as pioneering the modern style of magic that is still influential today (see Chapter 4). His approach was defined as much by what it rejected as what it embraced. Against the associations in the mid-1800s of magic with the mere entertainment of the street corner or the fairground, Robert-Houdin fashioned a persona of himself as a gentleman of society within the dignified space of an upmarket theater. As a showman, he took the conjuror's role as that of evoking a sense of wonder. In contrast to the extravagant props, clothing and scenery that characterized much of stage performances during his time, Robert-Houdin's stage set-up was minimalistic in appearance. Mechanical, optical and electrical gadgets that enabled his onstage effects were hidden from sight.[8] Wonder was generated through blending claims to astonishing powers with references to science, progress, and modernity. For instance, his 'Light and Heavy' trick was billed as an experimental demonstration of magical security. It employed a seemingly small wooden box that could both be lifted by a child and then somehow rendered immobile to the strongest adult.

As a performer then, Robert-Houdin combined the mannerism of the modern gentlemen with the inquisitiveness of a man of science, with the shrewdness of an illusionist. The contention that just such a dynamic was at play in his autobiography has been advanced by many scholars ever since. Perhaps the most notorious aspects of his 1858 memoir (*Confidences d'un Prestidigitateur*) relates to a trip to Algeria in 1856. After repeated invitations by the Political Office in Algiers, the conjuror tells of being brought to quell Arab anti-colonial resistance to French rule. Much of this resistance was attributed to a religious tribe called the Marabouts who claimed supernatural powers. In *Confidences*, Robert-Houdin recounts his performance of illusions intended to 'startle and even terrify' local Arabs 'by the display of a supernatural power'.[9] In

---

8   Smith, Wally. 2015. 'Technologies of Stage Magic: Simulation and Dissimulation', *Social Studies of Science*, 45(3): 319–343. https://doi.org/10.1177/0306312715577461.
9   Published in English as Robert-Houdin, Jean-Eugène. 1859. *Memoirs of Robert-Houdin. Ambassador, Author, and Conjurer*, R. Shelton Mackenzie (Ed.). Philadelphia: George G. Evans.

pursuit of this goal, he recounts performing the aforementioned 'Light and Heavy' trick beginning with the declaration:

> From what you have witnessed, you will attribute a supernatural power to me, and you are right. I will give you a new proof of my marvelous authority, by showing that I can deprive the most powerful man of his strength and restore it at my will. Anyone who thinks himself strong enough to try the experiment may draw near me.[10]

Not only was this version of the 'Light and Heavy' trick used to suggest the conjuror could control the native volunteer's strength through his supernatural power, but he also continued by shocking the volunteer through the metal of box handle to drive home his abilities.

As a portrayed confrontation of East-West as well as superstition-rationality, *Confidences* regales in what Robert-Houdin presented as his demystifying of primitive beliefs.[11] As anthropologist Graham Jones has argued, though, independent historical evidence for Robert-Houdin's claimed success in shifting local beliefs is scant. What evidence does exist suggests the Algerian audience regarded Robert-Houdin's performances as a form of entertainment rather than a convincing demonstration of the supernatural.[12]

Confidences also posits that, as a young journeyman learning his trade, Robert-Houdin chanced upon the travelling carriage of an aristocrat named Edmond de Grisy; a man that was also an expert magician with the stage name of Torrini. Over several chapters, *Confidences* recounts how Torrini cared for the young conjurer and took him under his wing as an apprentice.[13] While providing Robert-Houdin with a respectable lineage for what was typically regarded as a lowly art, what—if any—place this aristocrat had in the life of Robert-Houdin has come under much scrutiny.[14]

---

10 *Ibid.*
11 Leeder, Murray. 2010. 'M. Robert-Houdin Goes to Algeria', *Early Popular Visual Culture*, 8(2): 209–225. https://doi.org/10.1080/17460651003688113.
12 Jones, Graham. 2017. *Magic's Reason*. London: University of Chicago Press: Chapter 1.
13 Published in English as Robert-Houdin, Jean-Eugène. 1859. *Memoirs of Robert-Houdin. Ambassador, Author, and Conjurer*, R. Shelton Mackenzie (Ed.). Philadelphia: George G. Evans.
14 Metzner, Paul. 1998. *Crescendo of the Virtuoso: Spectacle, Skill, and Self-Promotion in Paris During the Age of Revolution*. London: University of California Press: Chapter 5.

However, not by all. Despite adopting a stage name derived after reading Robert-Houdin's autobiography, the escape artist that would become known as Harry Houdini (born as Erik Weisz), would later turn against his one-time inspiration. His 1908 book, *The Unmasking of Robert-Houdin*, attempted to dismiss the Frenchman's contribution to magic; indeed, the argument sought to expose the 'King of Conjurors' as tantamount to a fraudster. Yet, as magic historian Jim Steinmeyer details, while presenting himself as an authoritative historian of magic, ironically Houdini ended up taking the story of Torrini as genuine.[15]

Houdini's efforts to cement a place within the scholarship of magic went beyond criticism of prominent conjurors. Attempting to appropriate the authority of the encyclopedia, he offered entries on magic in the 1926 edition of the *Encyclopaedia Britannica* that placed himself as the central figure in magic. Mangan characterized Houdini's foray into encyclopedic writing in these terms:

> The magician's act depends upon such a sense: people go to see him precisely because of those special powers. And Houdini, of course, being the supreme myth-maker and self-publicist that he was, was hardly going to let an opportunity like writing the definitive encyclopaedia article slip by him. Because the very fact of the encyclopaedia's implicit claims of objectivity, authority and truthfulness work to his advantage: they provide a perfect kind of misdirection, a backdrop against which the textual performance of Harry Houdini can take place.[16]

A Reading of Autobiographies

The previous section surveyed some of the grounds for caution scholars have identified regarding the truth status of magicians' self-writing. Informed by Mangan's suggestion to treat such accounts as extensions of entertainment performances, the remainder of this chapter turns toward contemporary autobiographies.

Herein, though, attention proceeds with a deliberate tact. As suggested above, a common orientation to magicians' autobiographical accounts is to evaluate them through marshalling a sense of the

---

15 Steinmeyer, Jim. 2003. *Hiding the Elephant: How Magicians Invented the Impossible and Learned How to Disappear*. New York: Carroll and Graf: Chapter 7.
16 Mangan, Michael. 2007. *Performing Dark Arts: A Cultural History of Conjuring*. Bristol: Intellect: xxi.

truth—such as what historical records actually demonstrate. Through doing so, fact can be sifted from fiction, candor from duplicity, reality from appearances, etc.

Settling what's what, however, is not the goal of this chapter. Instead of reading the autobiographies like a historian seeking to establish the truth, I offer a reading of autobiographies as a student seeking to appreciate how performers perform. In particular, I examine how the aim of 'truth-telling' is and is not, made relevant by conjurors in their self-writing. As I will demonstrate, authors themselves can both anticipate and vindicate readers' skepticism about whether they are telling the truth. Not only this, authors can query whether telling the truth matters. This chapter aims to contrast how several leading magicians have positioned truth and deception in their accounts, and in doing so presented images of themselves, their audiences as well as their readers.

Penn & Teller: Playful Hustlers

Penn Jillette and Teller have provided prominent faces for American entertainment magic for decades. With their complementary mannerisms and appearances, the duo has garnered both considerable popular attention (including through their TV program *Fool Us*) and professional praise for their performance sophistication.[17] One dimension of that sophistication is how they artfully and selectively reveal their methods. Their first book, *Cruel Tricks for Dear Friends*, speaks to both how they pulled off some of their celebrated effects, as well as their autobiographical journeys into this tradecraft.

The matter of truth-telling is made relevant in *Cruel Tricks for Dear Friends* from the start. The book is about how to deceive. The first chapter, 'The No-Work, High-Yield, All-Electronic Computerized Card Trick' specifies step-by-step instructions for how to identify a randomly chosen card from a deck through calling a special telephone number set up by Penn & Teller. As they write, the:

---

17 For instance, see Neale, Robert E. 2008. 'Illusions About Illusions'. In: *Performing Magic on the Western Stage: From the Eighteenth Century to the Present*, Francesca Coppa, Lawrence Hass, and James Peck (Eds). London: Palgrave: 217–230.

...telephone number is hooked into a computer system which enables the tones of your Touch-Tone telephone to control a digitally recorded random-access compact disk. This permits you to do the ultimate card trick, in which virtually all the work of the magician is done with an electronic off-on flow-chart. To be sure you understand how to do the trick, do a dry run (without betting) on a trusted friend.[18]

As promised, after the seven-step dry run, the reader will be able to triumph over a 'sucker'.[19]

More than simply telling readers about how to scam, *Cruel Tricks for Dear Friends* plays with truth-telling through its physical construction. The book consists of three different kinds of printed pages. As the authors explain on page 108:

Did you notice these pages are a bitch to turn? [...] It wasn't your fault. The book is made that way. All of the pages are specially cut. If you play with this book a little, you'll notice that if you put your thumb on the edge and flip it front to back, all the pages look like itty bitty tiny irritating psycho-print with patterns printed over it, and if you flip from back-to-front it's all big stupid print.[20]

As the authors elaborate, these two formats, in addition to the third standard one, serve as the apparatus for pulling off an elaborate ruse designed to 'make a friend of yours look like a jerk'.[21]

Form and content come together in the disclosure on page 102 that all the 'attention-grabbing red instructions throughout this book are bogus. They are lies for you to use. Lies that will be your new friends'. Included within the considerable amount of red text in *Cruel Tricks for Dear Friends*, for instance, is the above quote for the 'The No-Work, High-Yield, All-Electronic Computerized Card Trick'. Indeed, almost all of the text for this trick is in red. In this way, directions *to* the readers about how to fool and deceive *others* are themselves eventually divulged to be instances of fooling and deception on the *reader*.

While some of the stories told in *Cruel Tricks for Dear Friends* are clearly or suggestively fictitious, others are written in a realist style. In the entry

---

18 Penn, Jillette and Teller, Raymond. 1989. *Cruel Tricks for Dear Friends*. New York: Villard Books: 4.
19 *Ibid.*: 4.
20 *Ibid.*: 108.
21 *Ibid.*

'The Scleral Shells', Teller recounts the back story behind an appearance on the early morning US television program *Today*. As part of the appearance, a *Today* presenter selected a card from a deck held by Penn. After Penn's failed attempt to identify the card, the suit and number of the card were shown to be written on Teller's eyes. The entry details the step-by-step procedures whereby the duo realized their initial idea, most notably visiting an ocular prosthetist to obtain sclera eye covers. The entry includes photos of Teller in the prosthetist's examination chair before, during and after the application of casting paste over his eyes. This crafted behind-the-scenes story ends with Teller's recollection of his response to the prosthetist's query about how the 'woman-sawed-in-half' trick gets done:

> A magician's sacred obligation is to keep the secrets of his brotherhood. Nothing brings about the ruination of the art form more quickly than low scum who betray their brethren and expose the methods of classic tricks.
>
> "Two women", I said. "One curls up in the head half of the box. The other is hidden in the tabletop and sticks her feet out when they are turning the box around. You don't notice the thickness of the tabletop because it's beveled. Anything else you want to know?"[22]

In this way, Teller exposes the mechanics of a trick, while calling into question those that expose tricks, as part of a matter-of-fact exposition of a trick.

Cruel Tricks for Dear Friends offers various revelations about the world of entertainment magic, Penn & Teller's performances of magic, as well as Penn and Teller as individuals. No simple claim to open disclosure is on offer, however. What is disclosed and what is yet concealed is a topic of explicit commentary, at least at certain points. Moreover, Penn & Teller admit they have lied in the past. But they go much further than this too; *Cruel Tricks for Dear Friends* includes statements that undermine Penn & Teller's trustworthiness and credibility. This takes place because they openly admit that they lie, they lack remorse about their deceptions, and they tell the hidden truths about their art.

Their follow-up book, *The Unpleasant Book of Penn & Teller or How to Play With Your Food*, likewise blends autobiography with 'how-to'

---

22 Penn, Jillette and Teller, Raymond. 1989. *Cruel Tricks for Dear Friends*. New York: Villard Books: 78.

explanations. The effects in question include past stage performances, but also pranks, gags and general mischief-making.

As part of these stories, again, what is on offer are claims whose truth status gets explicitly called into question. For instance, concerning telling one experience from his youth, Penn warns readers 'it's impossible to reconstruct the real facts, I've told this story so many times the real facts have disappeared'.[23] More pervasively in *How to Play With Your Food*, the authors repeatedly undermine their trustworthiness by explicitly rejoicing in how they scam the credulous for a living, how they seek to keep up a bad boy persona through breaking professional conventions, as well as how they encourage others to lie.

Such destabilizing efforts are combined with other conventional forms of narration that take what is written in the text and shown in the photos as unproblematic. As with *Cruel Tricks for Dear Friends*, readers are given step-by-step walkthroughs of prominent and lesser-known effects. Except for the odd reference to lewd details, left out because they are inappropriate for a 'family book', the orientation is repeatedly taken that readers are being presented with all of the pertinent information to know how effects were done.[24]

Such matter-of-fact orientations are given alongside more overtly playful presentations and reconstructions. For instance, one entry describes how to look like you are tying a cherry stem with your tongue, as Sherrilyn Fenn did on the cult TV show *Twin Peaks*. Within this entry, a close-up photograph is given of a mouth with a cherry stem sticking out, purported to be of Fenn. Turn over the page, however, and a photo is given of Penn with lipstick on and a cherry stem sticking out of his mouth. The caption provided states: 'Okay, so it's not really Sherrilyn Fenn. When photo rights get tough, the tough put on lipstick.'[25]

As part of explicitly questioning what is being disclosed, Penn & Teller distinguish different kinds of readers. For instance, the back cover states that the instructions on page 58 will enable readers to get back the cost of *How to Play With Your Food* from a single meal. In fulfilment of the claim, the two authors describe how to get others to pay for your

---

23 Penn, Jillette and Teller, Raymond. 1992. *The Unpleasant Book of Penn & Teller or How to Play with Your Food*. London: Pavilion: 14.
24 Ibid.: 94.
25 Ibid.: 56.

lunch in one entry. The entry includes photographic illustrations with descriptive captions. Within the course of the text, however, they warn:

> Read the directions, but pay no attention to the illustrations. All the illustrations (and captions) on the next two pages are bogus. They are intended to mislead semiliterate freeloaders who browse the book in a store and try to steal the valuable information you have paid for.[26]

The attention to who is paying for what is in line with much of the rest of the book. Whether to perform tricks and whether to divulge their secrets are decisions frequently pitched by Penn and Teller in terms of the monetary rewards on offer.[27]

And yet, despite the playful ways in which the extent of truth-telling is blatantly called into question or reduced to monetary calculations, on occasion, bright lines are drawn in *How to Play with your Food* about what counts as transgressive. As with *Cruel Tricks for Dear Friends*, for those that profit from some claimed actual psychic or supernatural powers—such as the ability to bend spoons with their minds—Penn & Teller offer scathing condemnations. While the authors offer step-by-step details of multiple ways to appear to be able to bend spoons with one's mind or to possess other extraordinary powers, they also suggest readers should disclose to audiences the trickery used to accomplish such feats...eventually, at least.

The previous paragraphs have offered a characterization of how truth-telling figures as a theme in Penn & Teller's writing. As writers, they demonstrate their abilities as performers not simply by recollecting past dissimulations, but by engaging in them as part of the books. Doing so successfully, however, amounts to traversing a tightrope. If their forms of deception were not detectable, then readers would not recognize their prowess as crafty storytellers. Conversely, if everything they wrote was regarded as pure fiction by readers, the books would likely be treated as nothing more than a flight of fantasy. Instead of either course, Penn & Teller opt for mixing kinds of telling in such a way that seems to necessitate that readers wonder about just what is going on.

In doing so, Penn & Teller do not just talk the talk about how concealing and divulging are skillfully employed in magic, they walk

---

26   *Ibid.*: 59.
27   See, for instance, ibid.: 200.

the walk through exemplifying their onstage performances in their writing.

Let us pull back from magic for a moment. While we might normally assume that people tell the truth, or at least what they believe to be true,[28] when it comes to someone that has admitted lying, such a starting assumption becomes more problematic. Yet just because a person lied on some occasion in the past does not mean they are lying now. Thus, the question of whether the truth is being told is something that needs to be worked out, again and again.

Michael Lynch and David Bogen examined how an admitted liar related to truth-telling, using the 1987 testimony of Oliver North at the Iran-Contra US Congressional Hearing. The truth status of North's testimony came up as a topic for consideration at the hearing by some Committee members as well as North himself, not least because by this point in time North had already admitted to misleading Congress. Congressional Committee members seeking to piece together how US officials secretly and illegally sold weapons to Iran in order to fund the Contras in Nicaragua were faced with a quandary in assessing North's testimony. To be taken as a credible witness, too, North had to pull off presenting himself on this occasion as a 'sincere liar'.[29] Lynch and Bogen characterized the tensions of North's testimony by arguing it set out these paradoxical contentions:

1. Lying is justified to prevent our adversaries from knowing our secrets.

2. Our adversaries have access to this very testimony.

3. I am not now lying. *And I really mean it, honest!*[30]

Such overt tensions about the truth status over those who have admitted lying are not uncommon. Michael D. Cohen, the former attorney for Donald J. Trump, testified before the Committee on Oversight and Reform of the House of Representatives in 2019 regarding its investigation

---

28 Grice, Paul. 1989. *Studies in the Way of Words*. Cambridge, MA: Harvard University Press.
29 Within magic, the notion of an 'honest liar' is commonplace, as in Measom, Tyler and Justin Weinstein. 2014. *An Honest Liar*. Left Turn Films.
30 Lynch, Michael and Bogen, David. 1996. *The Spectacle of History*. London: Duke University Press: 43.

into President Trump. As Cohen had pleaded guilty the previous year to eight counts including campaign finance violations, tax fraud and bank fraud, it is hardly surprising that whether and when he was telling the truth before the Committee were matters that both Cohen and the Committee members repeatedly revisited. His opening statement attempted to divorce this specific testimony from the backdrop of his previous criminal violations:

> For those who question my motives for being here today, I understand. I have lied, but I am not a liar. I have done bad things, but I am not a bad man. I have fixed things, but I am no longer your "fixer," Mr. Trump.[31]

While acknowledging his past misdeeds, Cohen made the case for the sincerity of his testimony by arguing:

> I am not a perfect man. I have done things I am not proud of, and I will live with the consequences of my actions for the rest of my life.
> But today, I get to decide the example I set for my children and how I attempt to change how history will remember me. I may not be able to change the past, but I can do right by the American people here today.[32]

Those questioning Cohen's allegations that the US President was a racist, a con man, a cheat and much else besides included Donald J. Trump. And yet, Trump did so in a manner that did not seek to completely cast doubt on Cohen's testimony or character. Responding to a reporter's question about the claims made against him by Cohen, Trump said:

> And he lied a lot, but it was very interesting because he didn't lie about one thing. He said, "No collusion with the Russia hoax." And I said, I wonder why he didn't just lie about that too, like he did about everything else. I mean, he lied about so many different things. And I was actually impressed that he didn't say, "Well, I think there was collusion for this reason or that." He didn't say that, he said, "No collusion," and I was a little impressed by that, frankly. He could've gone all out. He only went about 95 percent instead of 100 percent.[33]

---

31 Testimony of Cohen, Michael D. 2019. Committee on Oversight and Reform U.S. House of Representatives, February 27. https://www.theguardian.com/us-news/2019/feb/28/trump-says-cohen-lied-testimony-congress
32 *Ibid*.
33 Rupar, Aaron, 'Trump is "Impressed" that Cohen said "No Collusion." But Cohen Didn't Say that', *Vox*, 2019, February 28. https://www.vox.com/2019/2/28/18244483/trump-cohen-testimony-vietnam-news-conference-collusion

Herein, even for someone on the receiving end of damning allegations, whether an admitted liar is lying on a specific occasion is a matter presented as needing to be worked out.

To return to the accounts of conjurors, in contrast to attempts to draw clear boundaries around the truth, Penn & Teller engage in a much more varied and playful telling. At stake in this telling is both whether they are providing the truth and whether it matters.

While bright lines are set out at times about what counts as a transgression, falsehood, etc., Penn & Teller combine such appraisals with many other claims that offer grounds for doubting whether truth or lies are being told, to who and when, as well as how the tellers ought to be regarded. At points in their books, it matters whether the truth is being told; at other times this seems less relevant. Against the rampant humor and playfulness evident in *Cruel Tricks for Dear Friends* and *How to Play With Your Food*, an attempt by readers to get to the bottom of the truth status of what is written seems like a misplacement of energy at best. At worst, it is a profoundly misguided pursuit. Like an entertaining yarn, the question of what is going on seems much less important than enjoying the ride.

In short, then, the vision enabled by Penn & Teller is kaleidoscopic. One, but just one, of the kinds of argumentative contradictions they set out could be portrayed as:

1. Lying is justified to suckers that don't understand tricks;
2. You bought our book because you didn't know how to do tricks;
3. We are not now lying to you. *And we might not mean it, really!*

Derren Brown: An Authoritative Card

In the early 2000s, Derren Brown rose to notoriety in the United Kingdom through his television series *Derren Brown: Tricks of the Mind*. This series and many subsequent television and stage performances presented him as mixing 'magic, suggestion, psychology, misdirection and showmanship' as part of accomplishing baffling mental feats of prediction, mind reading, influence and much besides. Brown not only performed such deeds but occasionally discreetly taught members of the public how to do them.

In his 2007 book *Tricks of the Mind*, Brown offered a learned survey of a range of topics relevant for his tradecraft, including magic, hypnosis and unconscious communication. From the start of *Tricks of the Mind*, 'truth-telling' was made relevant. In its most blatant form, the book contains a sub-section early on entitled 'Truth and Lies'. As part of this, Brown raised what he characterized as the:

> rather embarrassing question of how honest I'm going to be with you when discussing my techniques. Some areas of the gutter press and of my own family seem convinced that amid the wealth of unmistakable candour, even-handedness, incorruptibility, rectitude and probity that has characterized my work to date, there might lie the occasional false or disingenuous datum designed to throw the careful seeker off course. Well, as my great-grandmother once said: rectitude and probity, my arse.[34]

The mix of truth-telling with humor that features in this excerpt features elsewhere in the book. As with Penn & Teller, Brown combines candid language with a nod and a wink, tongue-in-cheek style of writing that suggests readers ought to be on guard.

The 'Truth and Lies' sub-section concludes with Brown promising that: 'For reasons of space, practicality and retaining some mystery I cannot explain everything here; so in return for not being impossibly open, I promise to be entirely honest. All anecdotes are true, and all techniques are genuinely used'.[35] The importance of truth-telling is evident in other respects. From its front to back cover, *Tricks of the Mind* defines and debunks bad thinking and pseudo-scientific beliefs. The detailed exploration of the techniques and psychology of magic, hypnosis and much besides is not meant to 'make a friend of yours look like a jerk' or to hustle some money from an unsuspecting mark. Instead, *Tricks of the Mind* aims to provide a way of thinking clearly about often hazy topics. The list of topics ranges from Christianity to relativism to New Age spiritualism to environmentalism to alternative medicine.

However, more than simply being an effort to tell readers where the line exists between proper and improper thinking, *Tricks of the Mind* sets out instructions whereby the reader can *demonstrate to themselves* the psychological principles underlying Brown's performances. Want to see the power of suggestion? By just fashioning a rudimentary

---

34   Brown, Derren. 2007 *Tricks of the Mind*. London: Channel 4 Books: 14–15.
35   *Ibid.*, 19.

pendulum and following some brief instructions, readers can witness for themselves how objects can be made to move through the power of thought alone.

Such instructions have been taken as a straightforward 'how to' manual by some commentators on Brown.[36] However, it is possible to advance another way to interpret the instructions. This is the case because, at times, the try-it-yourself directions are overtly presented as limited disclosures too. The basics of ideomotor movement, hypnosis and card magic are elaborated, but the precise relationship between such descriptions and Brown's televised performances are often not drawn. Instead, readers are openly asked by Brown to take him on trust: for instance, in relation to how general explanations that Ouija boards rely on participants unconsciously moving its piece to the expected letters could account for how he used the same principle when the letters were *concealed* from participants.[37] As a result, method explanations are given in *Tricks of the Mind*, but some are highly limited in what they reveal.

Tricks of the Mind also places the seeds of doubt for what trustworthiness should be invested in Brown. In the 'Truth and Lies' sub-section, for instance, Brown both shared confidences about his past performances whilst noting that readers might have been duped by them, offered facetious self-boasts whilst stressing the need for self-deprecation from magicians, and portrayed honesty in magic as inherently problematic. Tongue-in-cheek portrayals of himself, his fans and the reader are abound in *Tricks of the Mind*.[38] Through such combinations, readers are positioned as needing to be able to distinguish for themselves what is actually meant from what is literally written.

One of the argumentative contradictions that could be derived from *Tricks of the Mind* is thus:

1. Distinguishing truth from falsehood is difficult as delusional thinking is rife;

---

36 Hill, Annette. 2010. *Paranormal Media: Audiences, Spirits, and Magic in Popular Culture*. London: Routledge: 142–149. https://doi.org/10.4324/9780203836392; and Mangan, Michael. 2017. 'Something Wicked: The Theatre of Derren Brown'. In: *Popular Performance*, Adam Ainsworth, Oliver Double and Louise Peacock (Eds). London: Bloomsbury: Chapter 6.
37 Brown, Derren. 2007. *Tricks of the Mind*. London: Channel 4 Books: 48.
38 For examples, see pages xv, xvi, 5 and 7.

2. My work has sought to tap into delusional forms of thinking to persuade and deceive;

3. You can tell for yourself what is true from what I am telling you now, *and I mean it, really!*[39]

Contrast, then, *Tricks of the Mind* with a second autobiographical book written for the general public by Brown in 2010 entitled *Confessions of a Conjuror*. Whereas the former reviews magic, hypnosis, memory and unconscious communication through drawing on his experiences, the latter book squarely starts from autobiographical experiences to speak to the shared fallibility of our minds. In *Confessions of a Conjuror* Brown presents himself as susceptible to flawed ways of reasoning: snap judgements, blinkered perceptions, confirmation seeking and so on. Even as a professional conjuror with years of experience in befuddling audiences, his thinking is presented as imbued with chains of personal associations and questionable lines of inference. Where does his dislike of blue playing cards come from? Why does he feel the need to impress

---

39 Some commentators on magic have identified parallels in Brown's performances to the ambiguities identified above in *Tricks of the Mind*. Whilst seeking to counter belief in spiritualism, superstition and much more, in his television programs Brown has provided more or less elaborate explanations for his feats related to principles of psychology, hypnotism and subliminal messaging. Yet, as some have argued, such lines of explanation have themselves served as misleading forms of misdirection that work to distract audiences from the true methods employed. (For a discussion of the ambiguity sought, see Brown, Derren and Swiss, Jamy Ian. 2003, June 29. *A Conversation in Two Parts: Part I*. Available at http://honestliar.com/fm/works/derren-brown.html). In placing a scientific explanatory gloss on effects achieved with 'little more than clever magic tricks' (Singh, Simon. 2003, June 10. 'I'll Bet £1,000 That Derren Can't Read my Mind', *The Daily Telegraph*), the criticism levelled at Brown has been that he too has promulgated pseudo-beliefs (see as well *Magic, Charlatanry and Skepticism*, SOMA Workshop. https://scienceofmagicassoc.org/blog/2021/4/29/magic-charlatanry-skepticism-webinar-cd6cy). In response, in his more recent work, Brown has noted to audiences how his scientific glosses of tricks have served as a means of ruse (see Mangan, Michael. 2017. 'Something Wicked: The Theatre of Derren Brown'. In *Popular Performance*, Adam Ainsworth, Oliver Double, and Louise Peacock (Eds). London: Bloomsbury: Chapter 6). Yet another line of criticism that has been made against him, however, is that even when he exposes his own pseudo-explanations, the actual effects of his performances on audiences might be to reinforce pseudo-beliefs (see Malvern, Jack. 2019, January 2. 'Magicians Accused of Casting Pseudoscience Spell on Audiences', *The Times*, and for a more general analysis of these issues for magic overall see Jay, J. 2016. 'What do Audiences Really Think?' *MAGIC* (September): 46–55. https://www.magicconvention.com/wp-content/uploads/2017/08/Survey.pdf).

his family and the famous? Why does he keep falling for crass sales techniques?

Yet reasons can be offered for thinking that the extent of self-reflection is limited. Even as *Confessions of a Conjuror* 'invites you on a whimsical journey through his unusual mind'[40] in a manner that emphasizes self-admission, in other respects the degree of disclosure seems to have been decidedly pushed to the margins. For instance, *Confessions of a Conjuror* is structured through Brown recounting an extended card effect he performed for a group at a restaurant table in Bristol. A not-insignificant amount of the 327 pages of the paperback version is dedicated to a clear-cut description of the minutiae of the encounter: the moment-to-moment subtle and gross verbal and non-verbal expressions of the three spectators; the precise mechanics of his performance in response to their expressions and actions; the train of mental associations launched in his mind at the time; the physical details of the scene, and so on. Certainly, in reading this extent of detail about this one encounter, I was left wondering how the exactitude and degree of disclosure could be possible. This was especially so given the repeated contention in *Confessions of a Conjuror* that human recall is defective.

Layered Truths

The autobiographical accounts surveyed so far in this chapter have identified, named and organized events and experiences to present an understanding of authors, their crafts and the world. As with other types of memoirs, and indeed much of what gets told secondhand in everyday life, evaluating the reliability of the claims given is challenging. As authors recount events for which few readers will have any direct knowledge, concerns that spinning, slanting and the like might be at play in authors' portrayals of themselves cannot be completely quashed.

In this chapter, I have proposed that, in the case of modern conjuring, such underlying grounds for doubt are complemented by specific concerns about magicians as authors. Since their credibility as practitioners in part derives from their ability to skillfully deceive, dissimulate and simulate, readers have specific reasons to wonder about

---

40 As on the back cover of Brown, Derren. *Confessions of a Conjuror*. London: Channel 4 Books.

what conjuror-authors are getting up to in their writing. In recognition of the justification for suspicion by readers, authors such as Penn, Teller and Brown actively pulled back (at times) from treating their accounts at face value. Such pulling back itself was positioned as a marker of their skill, not a refutation of it. In short, these three authors portrayed themselves as credible authors not by overlooking potential concerns about their reliability and genuineness, but by acknowledging and finding some way of working with suspicions. At least, that is, to some extent.

Not all prominent figures in magic have opted to explicitly cast doubt on their truth-telling. Within the pages of *Nothing Is Impossible*, the British magician Dynamo gives readers an off-stage account of his meteoric rise to fame.[41] His life story includes a description of the struggles he experienced growing up, the influences on his magic, as well as the hard graft of becoming known. The style is one of opening up to the reader about his life. At times though, too, it is clear that Dynamo is not being fully forthcoming. Especially relating to how he accomplishes his effects, details are scant. For instance, in describing his feat of walking down the side of the building of the newspaper the *Los Angeles Times*, he writes 'I knew that if my magic didn't work, there was no way I would survive the fall'.[42]

And yet, besides these discretions surrounding methods, there is little by the way of overt acknowledgement of the need for caution about what is printed. Instead of drawing attention to the limits of his trustworthiness as an author, *Nothing Is Impossible* brings readers backstage to witness the unappreciated story behind Dynamo—just as it happened.

In his later 2017 book *Dynamo: The Book of Secrets*, Dynamo offers detailed instructions for dozens of effects whilst also providing autobiographical insights. Rather than advocating scamming others or debunking those that scam, *Dynamo: The Book of Secrets* sets itself out as enabling readers to emotionally affect people through playing cards, rings, pens and other everyday objects.[43] As with *Nothing Is Impossible*, not only then does the truth matter, the instructions and the

---

41 Dynamo. 2012. *Nothing Is Impossible*. London: Ebury Press.
42 Ibid.: xx.
43 Dynamo. 2017. *Dynamo: The Book of Secrets*. London: Blink Publishing.

autobiographical notes are treated as placing truthful insights before the reader. In this way, Nothing Is Impossible offers a mirror to reality.

And yet, though distrust is not treated as warranted, neither is everything presented at face value. *Dynamo: The Book of Secrets* presents itself as layered in a play of secrecy. As set out in the introduction, while 30 powerful effects are sketched, 'if you read between the lines, there are even more secrets to uncover'. The latter part of this sentence is underlined with an arrow leading away to a text written in a different font stating 'I'm serious about this'.[44] With such an invitation, instead of casting doubt on the sincerity of the contents along the lines of Penn & Teller in The No-Work, High-Yield, All-Electronic Computerized Card Trick, readers of *Dynamo: The Book of Secrets* are encouraged to hunt for even deeper truthful confidences hidden within the text. Evoking another, unelaborated level of secrets not only reinforces the veracity of what is presented. Delving deeper provides a basis for some (diligent) readers to set themselves apart from the (surface level) sense-making of more casual readers.[45]

With the purported employment of this literary technique of layered truths, Dynamo adopts the position of someone skilled in mixing disclosure with concealment. Such a layered presentation of truths is evident elsewhere. David Blaine first rose to international prominence in the late 1990s with television shows such as *David Blaine: Street Magic* and then later for his feats of endurance. His first book, *Mysterious Stranger*, mixes 'how-to' instructions for readers with recollections from his early and later years. What is it like to be immersed in a block of ice or encased underground? How do you convince others you are psychic? *Mysterious Stranger* provides answers to such questions and does so without supplying overt grounds for doubting the trustworthiness of Blaine or the literalness of what is presented.

The adequacy of how this was done is open to question. For instance, in *Mysterious Stranger*, Blaine's career trajectory is recounted through situating his performances within the traditions of gurus, escape artists, pillar saints, con artists and the like. In doing so, Blaine placed much credence in the abilities of such individuals and regaled in their claimed exploits. And yet, whether this overall deference is warranted might

---

44 Ibid.: 10.
45 And, it seems worth noting, I never found any deeper secrets.

well be doubted. Robert-Houdin's claims regarding the effects of his magic show on the native uprising in Algeria noted above, for instance, are repeated by Blaine without critical scrutiny or recognition of their contentious status within the history of magic.[46] Also concerning truth-telling, Blaine embraces actions that other magicians question. Not only are financial scams told without moral condemnation, but Blaine also advises readers how they can play on (rather than dispel) audiences' superstitions.[47]

Layered meaning comes into *Mysterious Stranger* but in a different way than in *Dynamo: The Book of Secrets*. *Mysterious Stranger* is presented as a puzzle. As printed on the jacket cover, 'Hidden throughout the Book are secret signals, codes and clues, that once understood and deciphered will lead to the discovery of a TREASURE which has been hidden somewhere in the United States of America'.[48] Blaine offered a reward of $100,000 to anyone that could find the treasure. In this way, *Mysterious Stranger* echoed the long history of magicians advertising substantial prizes to those able to discern their hidden secrets.[49]

Legit Cave

Previous sections of this chapter have provided my reading of the diversity of stances prominent magicians have taken regarding truth-telling. In adopting such stances, the autobiographies of modern conjurors cut a complex relation to established genres of writing involving truth-telling about the self. They are not simply confessionals. As contended by Roth and De Man, confessionals are typically presented as motivated by feelings of guilt and shame over one's misdeeds.[50] Despite the forms of deception involved in magic, such inner sentiments of shame were not aired by Penn, Teller, Brown, Blaine, Dynamo or Robert-Houdin. As

---

46 For instance, see Blaine, David. 2002. *Mysterious Stranger*. New York: Villard: 146.
47 Blaine, David. 2002. *Mysterious Stranger*. New York: Villard: 16 and 34.
48 *Ibid*.
49 Practices not completely divorced from the aims of garnering publicity as well as promoting performers' repute.
50 Roth, Ben. 2012. 'Confessions, Excuses, and the Storytelling Self: Rereading Rousseau with Paul de Man'. In: *Re-thinking European Politics and History* (Vol. 32), A. Pasieka, D. Petruccelli, B. Roth (Eds). Vienna: IWM Junior Visiting Fellows' Conferences; and De Man, Paul. 1979. *Allegories of Reading*. New Haven: Yale University Press.

a rule, magicians do not express shame over their efforts to misdirect, deceive or dissimulate.

Moreover, though, confessions typically assume that the truth can be definitely established (or, at least, the truth as understood by the writers). At least some of the authors surveyed in this chapter, though, labored to sow doubt regarding whether they were providing a straight account.

But if the autobiographies are not well characterized as confessionals, they sit uneasily with the alternative label of excuses. Excuses are attempts to explain away responsibility for behavior that is likely to be regarded as transgressive.[51] As elaborated above, the authors surveyed rarely admit to engaging in questionable moral conduct. When they do so, as in the case of Penn & Teller, they often revel in the transgression rather than try to excuse it.

Justifications are more common than excuses. Herein, authors accept responsibility for their actions, but offer grounds as for why those actions should not count as morally dubious. Dynamo, for instance, justified why he was able to tell magic's secrets in *Dynamo: The Book of Secrets*. In line with a refrain that figured within his television series in the 2010s, early on in the book he underscores the importance of keeping secrets. This is done through recounting a formative experience when his grandfather performed an enchanting piece of magic on him that he states he still cannot figure out today. Against these considerations though, the introduction sets out why he is correct in telling secrets in *Dynamo: The Book of Secrets*:

> Magicians aren't supposed to reveal their secrets, right? That's true—I keep many of the effects in my repertoire so secret that I haven't told anyone how I do them. But the pieces in this book are different. I have picked effects that are perfect for people new to magic to learn because they are easy to do but get reactions. There is a world of difference between teaching magic and exposing it and I am teaching these effects because they are the perfect starting point to a life (or even a career) in magic.[52]

---

51 De Man, Paul. 1979. *Allegories of Reading*. New Haven: Yale University Press.
52 Dynamo. 2017. *Dynamo: The Book of Secrets*. London: Blink Publishing: 11.

In this way, the secrets of magic can be told because the effects themselves are those for beginners and the purpose of telling is to teach, not to divulge.

Instead of making a plea for absolution, Penn & Teller adopt multiple orientations to truth-telling that invite readers to speculate about the ultimate truth status of what is being told, as well as whether their inner selves are being revealed. At times they embrace the certainty that they have done wrong and they most certainly *are* blameworthy. Brown, as well (at least in certain respects) overtly acknowledges the artfulness of his self-presentation. Through doing so, these authors cut against the grain of the play for authenticity that typically characterizes self-narratives.[53] Whatever might be lost in term of authenticity, however, the mix of truth and deception is presented as enhancing their credibility. Penn, Teller and Brown display playfulness and wit to the readers. In talking the talk, they walk the walk regarding their ability to simulate and dissimulate.

Through surveying what we might call the 'skilled revelation of skilled concealment'[54] that constitutes the autobiographies of leading conjurors, my aim in this chapter has been to convey learnt sensitivities associated with how truth and deception are alternatively marshalled within one genre of writing. Within this, one sub-goal has been to explore how motion and mix—in what is laid before and what is still occulted away, as well as the way 'things are what they seem' and 'things are not what they seem'—*together* create autobiographical revelations. That motion and mix can extend to the identity of the authors, who can varyingly be presented as trustworthy, reliable, stable, elusive and so on. Motion and mix also extend to the image of readers. As contended, readers of autobiographies are varyingly told that we are being brought into the fraternity of those in on a gag; we are being offered a partial glimpse through a keyhole, and we should wonder whether we have been left out in the cold.[55]

---

53 Atkinson, Paul and Silverman, David. 1997. 'Kundera's Immortality', *Qualitative Inquiry*, 3: 304–325.

54 To borrow a phrase from Taussig, Michael. 2016. 'Viscerality, Faith, and Skepticism', *Hau: Journal of Ethnographic Theory*, 6(3): 455. https://doi.org/10.14318/hau6.3.033.

55 The arguments of this chapter might well encourage some readers to reflect on how I as an author-magician have been crafted in this book. For an example of how I as author-magician engaged in such forms of playful writing, see Rappert, Brian. 2021. 'Conjuring Imposters'. In: *The Imposter as Social Theory*, Steve Woolgar, Else Vogel, David Moats and Claes-Fredrick Helgesson (Eds). Bristol: Bristol University Press: 147–170.

7. Control and Care

'You're not playing with us, you are playing on us.'

– Anonymous

'We don't keep secrets *from* the audience, we keep secrets *for* the audience.'

– Michael Weber[1]

How should we be together?

In late 2019, as part of my apprenticeship in conjuring, I began undertaking paid-for small group shows akin in their basic format to the sessions discussed in Chapter 3.[2] At each show, a dozen or so participants assembled around a large table at the Ashburton Arts Centre near the edge of Dartmoor National Park. A series of magical effects were interspersed with group discussion which I prompted and then moderated. Eight events were held before the Covid-19 lockdown in England. After lockdown began, the sessions moved online, with 16 held through the Ashburton Arts Centre and the Exeter Phoenix up until February 2021.

Toward the end of my first show in November 2019, I suggested to participants how magic can entail a playful chemistry of concealment and revelation. At this point, one person interjected with the comment at the start of this chapter. Introducing himself as a retired schoolteacher, he contrasted the open-ended way children can play with the orchestrated actions that made up the show. The exchange that followed was one of several memorable episodes for me, in which disquiet was openly voiced about how we came together.[3]

---

1 https://tomdup.wordpress.com/tag/michael-weber/
2 All public shows were held as charity fundraisers.
3 For a discussion on 'dark' forms of play in magic, see Dean, E. 2018. 'The End of Mindreading', *Journal of Performance Magic*, 5(1). https://doi.org/10.5920/jpm.2018.04

As a further way into understanding conjuring as a form of interaction, in this chapter I hold together the notions of 'control' and 'care' to ask how each can inform the other. As in previous chapters, I do so by surveying the thoughts of professional magicians, as well as reflecting on my experiences. On the latter, in starting this self-other study in late 2017, I had no sense that care would figure as a theme in my research. At that point, questions about how concealment and revelation mixed were at the forefront of my mind. By the time I began offering public shows, however, matters of care had become central. This chapter shares some of the sensitivities and strategies that emerged.

Control and Connection

As developed in earlier chapters, while conjuring is often theorized as an activity involving magicians and audiences, a tendency is to treat the encounter in one-directional terms. Certainly, when it comes to the magic effects themselves, control is often characterized as essential. Whilst conjuring is acknowledged as an activity done for an audience, frequently agency, knowledge and the scope for action is invested with the conjuror. Or, at least these things should rest there if conjurors are doing their jobs properly. To offer an analogy, conjuring is often likened to sculpting. Through skills of misdirection and much besides, the conjuror molds audiences' perceptions and understandings. Some audiences are rough, some pliable and some strained in their composition. The task of the conjuror is to achieve the desired effect against whatever niggles might present themselves.

And yet, while the imperative for control reoccurs in instructional materials and professional discussions, so too does the importance of making an emotional connection to others.[4] A frequent refrain is that magic is created with, not just for, the audience.[5] As a result, thinking about how to guard against belittling audiences or inflating one's

---

4   As, for instance, in Burger, Eugene n.d. *Audience Involvement…A Lecture* Asheville, NC: Excelsior!! Productions; as well as Vincent, Michael. 2021. Share Magic Lecture, 27 October. https://www.sharemagic.org/sharemagic/?utm_campaign=Michael+Vincent+ShareMagic&utm_content=Michael+Vincent+Share+Magic&utm_medium=newsletter&utm_source=email

5   Clifford, Peter. 2020, January 12. *A Story for Performance*. Lecture notes from presentation at The Session. London.

self-importance is a recurring concern.[6] Humor, storytelling and self-effacement are some of the techniques advocated to avoid appearing smug, superior and so on to audiences, even as they are subject to calculated manipulations.[7] As part of an explicit attempt to move away from conceiving of magic in one-directional terms of domination, Jon Allen spoke to a variety of techniques for seeking emotional resonance in his instructional DVD titled *Connection*:

1. Using physical props that people attribute with significance or can be made significant;[8]
2. Asking questions of audiences that can inform the magic;
3. Using meaningful themes and symbols (for instance, togetherness);
4. Having a personality;
5. Matching the energy of the audience;
6. Ensuring audience members interact;
7. Making sure everyone present participates and feels positive from the experience;
8. Being okay with struggling in front of audiences.[9]

Through undertaking these kinds of techniques, the objective[10] is for magicians to be with others even if a sharp rift exists between the spectators' and magicians' understandings of what is taking place.

---

6   For instance, Close, Michael. [2003] 2013. 'The Big Lie'. In: *Magic in Mind: Essential Essays for Magicians*, Joshua Jay (Ed.). Sacramento: Vanishing Inc: 97–102.
7   For instance, Nelms, Henning. [1969] 2000. *Magic and Showmanship*. Mineola, NY: Dover.
8   Objects can be imbued with significance for many reasons. For instance, as part of my online shows, audience members were asked to have a deck of cards to hand in order to do some effects together. In one case, a participant had gone into her attic to find the box containing magic tricks that her recently deceased father had used on special family occasions. As she discovered, the deck of playing cards he used was almost completely made up of the same identical card, the King of Diamonds.
9   Allen, Jon. 2013. *Connection*. Las Vegas, NV: Penguin Magic.
10 Whether or not the techniques spoken to in this section of the book work in the manner expected is another matter. For instance, evidence does exist suggesting that some audiences decidedly like to be fooled and most would rather observe than partake in tricks, see Jay, J. 2016. 'What Do Audiences Really Think?', *MAGIC* (September): 46–55. https://www.magicconvention.com/wp-content/uploads/2017/08/Survey.pdf

Other advice given by experienced professionals includes telling stories,[11] conveying messages,[12] giving gifts to audiences,[13] and even making tricks peripheral features of shows.[14] Whatever the specific technique, those seeking to realize an empathetic connection call for moving away from the tendency of conjurors to present magic as a puzzle-solving exercise that challenges audiences to figure out how 'tricks'[15] are done.[16] While magicians' fascination with technique makes puzzle-solving a suitable aim for when they perform for each other, creating astonishment for laypeople is said to require emotional engagement.[17]

The call for magicians to be responsive to audiences' predilections (rather than their own) indicates one limit to conceiving of conjuring as an exercise in unilateral control. Even as conjurors seek to influence how others behave and what they perceive, they must do so in ways that audiences regard as recognizable, intelligible and appropriate. Thus, magicians need to give up pursuing their likings and ensure they act in ways regarded as suitable for specific audiences.[18] These preferences express culturally available, historically formed and locally enacted conventions, mores and judgements. As Steve Palmore maintained, as an African-American magician often performing for Caucasian groups, the demands of meeting audiences' expectations can extend far beyond the minutiae of the naturality of hand movements.[19] Instead,

---

11   Neale, Robert. 1991. *Tricks of the Imagination*. Seattle: Hermetic Press and Jones, G. 2012. 'Magic with a Message', *Cultural Anthropology*, 27(2): 193–214. https://doi.org/10.1111/j.1548-1360.2012.01140.x

12   Allen, Jonathan and O'Reilly, Sally. 2009. *Magic Show*. London: Hayward Publishing: 84.

13   Hass, Lawrence. (Ed.) 2010. *Gift Magic: Performances That Leave People with a Souvenir*. Theory and Art of Magic Press.

14   Brown, Derren. 2021. *Bristol Society of Magic—Centenary Celebration: An Evening with Derren Brown* (Bristol), 3 May.

15   Along these lines, within performances magicians often avoid the language of 'tricks' and 'trickery' because of their connotations about fooling, in favor of other terminology for their feats such as 'event', 'experience', 'experiment' or simply 'something mysterious'.

16   Compare and contrast, for instance, McCabe, Pete. 2017. *Scripting Magic*. London: Vanishing Inc: 306; and Bruns, L. C. and Zompetti, J. P. 2014. 'The Rhetorical Goddess: A Feminist Perspective on Women in Magic', *Journal of Performance Magic*, 2(1). https://doi.org/10.5920/jpm.2014.218

17   As in Harris, Paul and Mead, Eric. *The Art of Astonishing*. [n.p.]: Multimedia A-1.

18   Comments by Paul Draper in 'Scripting Magic 2.1 (Part 2)', 11 September 2020. https://videochatmagic.substack.com/p/scripting-magic-21-part-2

19   Palmore, Steve. 2020. *Vanish*, 31: 25.

pervasive cultural stereotypes can come into play.[20] It is by successfully engaging with others' expectations and desires, not simply their own, that magicians gain validation from audiences that they are competent, convincing and charismatic.[21] Furthermore, even as magicians act to deceive others, doing so requires they labor to 'induce or suppress feeling so to sustain the outward countenance that produces the proper state of mind in others'.[22] Thus, conjurors need to police themselves for displays of emotion that might be deemed 'out of place' (and much more besides—see Chapter 4).

In sum, to do *for* another can entail doing *according to* another. And in commercial magic, there can be many others—audience members, performance venues, production houses, technical crews, online platforms, video directors, 'the market', etc.[23]

More than this, in attempting to exert control over others, magicians invariably make themselves vulnerable. While some conjurors might conceive of their central task as producing docility,[24] the prospect that others (for instance, children or partygoers) might not go along with such plans can never be fully eliminated. Indeed, it is the ability to go on in light of the possibility that things may go awry (because people and objects are not fully controllable) that, in significant respects, makes magic a skillful enterprise. The heckler, for instance, represents one, much professionally derided incarnation of unruliness.[25] This status

---

20 For a gender-based analysis of such issues, see Bruns, L. C. and Zompetti, J. P. 2014. 'The Rhetorical Goddess: A Feminist Perspective on Women in Magic', *Journal of Performance Magic*, 2(1). https://doi.org/10.5920/jpm.2014.218 and Noyes, P. and Pallenberg, H. 2008. *Women in Boxes: The Documentary Film About Magic's Better Half* [Motion Picture]. Available from http://www.filmbaby.com/films/3277

21 When magic is performed across cultures, what counts as magic skill can be much contested; see Goto-Jones, Chris. 2016. *Conjuring Asia: Magic, Orientalism, and the Making of the Modern World*. Cambridge: Cambridge University Press. https://doi.org/10.1017/cbo9781139924573.

22 Hochschild, A. 1983. *The Managed Heart: Commercialization of Human Feeling*. Berkeley: University of California Press: 7.

23 For a discussion of such matters, listen to Shezam. 2020. *Podcast 54—Catie Osborn on Shakespeare and Tips From an Entertainment Director*. Shezam Podcast. Available at https://shezampod.com/series/shezam/ and Frisch, Ian. 2019. *Magic Is Dead: My Journey into the World's Most Secretive Society of Magicians*. New York: Dey St.: 102.

24 For a detailed analysis on how this can be done in unfolding interactions, see Jones, Graham M. and Shweder, Lauren. 2003. 'The Performance of Illusion and Illusionary Performatives', *Journal of Linguistic Anthropology*, 31(1): 51–70.

25 Jones, Graham M. 2017. *Magic's Reason*. London: University of Chicago Press: 156.

stems, at least in part, from the manner that hecklers do not subscribe to the same notions as others who are present about who can speak when, about what, to whom and in what manner. They want to be 'IT, no matter what anyone else thinks'.[26] It is perhaps not surprising, then, that the handling of hecklers is portrayed as a vital and nuanced proficiency.[27] Maybe even worse than being heckled, though, audiences can simply leave, never to return. And without an audience, can there be magic? Or even a magician?

The Ethics of Care

The arguments in the previous paragraphs suggest reasons why control is not well understood simply as the command of the conjurer. To foster other ways of understanding, in the remainder of the chapter I approach magic through an alternative (yet not simply opposing) concept. Specifically, born out of the sensitivities fostered through my self-other study, I want to ask what comes to the fore when magic is approached as a *practice of care*.

To seek to care is to be motivated to think and act in relation to one's or others' needs. Attempts to theorize care, particularly developed within feminist studies, have led to varied formulations.[28] Although multiply conceived, care is frequently regarded as a practice of attention. More than just attention, caring has been said to entail a willingness to respond to needs, a competency in doing so and a regard for the possibility that responses can turn abusive.[29]

In recent decades, the concept of care has served as a basis for reimagining many relationships. How students and teachers, clients and professionals, as well as patients and doctors can be with one other

---

26   Hopkins, Charles. 1978. *Outs, Precautions and Challenges for Ambitious Card Workers*. Calgary: Micky Hades: 76.

27   See Nelms, Henning. 2000 [1969]. *Magic and Showmanship*. Mineola, NY: Dover: 232 and Jones, Graham and Shweder, Lauren. 2003. 'The Performance of Illusion and Illusionary Performatives: Learning the Language of Theatrical Magic', *Journal of Linguistic Anthropology*, 13(1): 51–70.

28   Gilligan, Carol. 1982. *In a Different Voice*. Cambridge, MA: Harvard University Press; Held, Virginia. 1993. *Feminist Morality: Transforming Culture, Society, and Politics*. Chicago, IL: University of Chicago Press; and Kittay, Eva Feder. 1999. *Love's Labor*. London: Routledge.

29   Tronto, Joan. 1994. *Moral Boundaries*. London: Routledge.

has been informed by the 'ethics of care'.³⁰ Rather than the asymmetries in such relations leading to one-sided conceptions of what it means to deliver care, though, the asymmetries have been taken as the basis for underscoring mutual dependency. This is so because the one who is cared for and the one caring realize themselves through each other. Caring cannot take place when those cared for reject what is offered or when carers disengage from the cared-for.³¹ Conceived as such, caring is a deeply ethical endeavor featuring vulnerability, responsibility and mutuality between all present.

With the centrality given to vulnerability, responsibility and mutuality, many of those theorizing care have asked how it can entail its notional opposites. For Aryn Martin and colleagues: 'Care is an affectively charged and selective mode of attention that directs action, affection, or concern at something, and in effect, it draws attention away from other things'.³² As they also argue, since our efforts and energies cannot be directed everywhere and to everyone, care:

> circumscribes and cherishes some things, lives, or phenomena as its objects. In the process, it excludes others. Practices of care are always shot through with asymmetrical power relations: who has the power to care? Who has the power to define what counts as care and how it should be administered?³³

Conceived in this manner, caring is a practice dependent on control.³⁴

With the ways attention and inattention are bound together, the question of whether 'caring' is taking place cannot be assumed from the outset. A hospital might be a quintessential site for care, but just because some people are visibly attending to others does not mean that 'caring' is an apt description for what is going on. Caring requires continually

---

30 See e.g., Reiter, Sara. 1997. 'The Ethics of Care and New Paradigms for Accounting Practice', *Accounting, Auditing & Accountability Journal*, 10(3): 299–324. https://doi.org/10.1108/09513579710178098
31 Noddings, Nel. 2013. *Caring: A Relational Approach to Ethics and Moral Education* (Second edition, updated). London: University of California Press.
32 Martin, Arrn, Myers, Natasha and Viseu, Ana. 2015. 'The Politics of Care in Technoscience', *Social Studies of Science*, 45(5): 635. https://doi.org/10.1177/0306312715602073
33 *Ibid.*, 627.
34 Pettersen, Tove. 2011. 'The Ethics of Care: Normative Structures and Empirical Implications', *Health Care Analysis*, 19(1): 51–64 https://doi.org/10.1007/s10728-010-0163-7

posing questions about how and why caring takes place, what it means to be receptive to others, how the cared-for contribute to caring, who can care in the first place and who defines what the term means.[35] In this way, its realization is often contrasted with simply going through the motions of assisting others. Caring is done in specific situations in which the question of how to act cannot be pre-determined. An ongoing 'openness concerning the very questions of what is cared for, how to care and who cares'[36] has been advised to prevent care from descending into carelessness.

Entanglements of Care

Let us return to conjuring. As suggested above, though rarely using the term care, magicians have questioned how they can be attentive and responsive to the emotional needs of their audiences. Beyond those already mentioned, additional entanglements can be noted regarding what counts as care, how to care and who should care in magic.

The place of deception is one source of unease. Conjurors routinely act with the intent to mislead. They strive to deceive through deliberate acts of dissimulation and simulation.[37] In this, they are hardly alone as professionals.[38] But still, the centrality of deception and secrecy in conjuring might be taken as in conflict with the possibility for responsiveness to others.

---

35 Hendriks, Ruud. 2012. 'Tackling Indifference—Clowning, Dementia, and the Articulation of a Sensitive Body', *Medical Anthropology*, 31(6): 459–476. https://doi.org/10.1080/01459740.2012.674991; and Johns, Christopher. 2009. *Becoming a Reflective Practitioner* (Third Edition). Oxford: Wiley-Blackwell.

36 Schillmeier, Michael. 2017. 'The Cosmopolitics of Situated Care', *The Sociological Review Monographs*, 65(2): 58. https://doi.org/10.1177/0081176917710426.

37 In a wide-ranging analysis of deception (including magic), Barton Whaley defined *dissimulation* as hiding the real, whereas simulation is showing the false. See Whaley, Barton. 1982. 'Toward a General Theory of Deception', *The Journal of Strategic Studies*, 5(1): 178–192. https://doi.org/10.1080/01402398208437106

38 Tuckett, A. 1988. 'Bending the Truth: Professionals Narratives about Lying and Deception in Nursing Practice', *International Journal of Nursing Studies*, 35(5): 292–302; Clarke, S. 1999. 'Justifying Deception in Social Science Research', *Journal of Applied Philosophy*, 16(2): 151–166. https://doi.org/10.1111/1468-5930.00117; and Hunt, Jennifer and Manning, Peter K. 1991. 'The Social Context of Police Lying', *Symbolic Interaction*, 14(1): 51–70. https://doi.org/10.1525/si.1991.14.1.51.

The quote from Mike Weber at the start of the chapter, however, provides one justification for secrecy.[39] Through obscuring the mundane methods for effects, audiences can experience wonder, astonishment and much else besides.[40] A parallel argument could be given for deception. And yet, even if secret-keeping and deception are taken as integral to inducing wonder, a countervailing danger is that the motivations for them can have more to do with bolstering the aura of magicians. In *The Royal Road to Card Magic*, for instance, Hugard and Braué propose the rule:

> Never reveal the secret of a trick. Many good card tricks are so simple that to reveal the method is to lower yourself in the estimation of the audience, who have given you great credit for a skill that you then confess you don't possess.[41]

Herein, it is the status of the magician that takes center stage. However, bolstering the standing of magicians can, at times, also be an act of caring. For instance, hospitalized patients have been taught magic as a way of fostering a sense of control in order to counter feelings of disempowerment.[42]

Doubleness characterizes other aspects of magic. For instance, in conjuring, as in social life more generally, one way to build a connection with another person is to visibly attend to them. Making eye contact and closely watching others' reactions are essentials for being responsive. And yet, the appropriateness of the magician's gaze at the culmination of a trick has been called into question.[43] The root of the concern is that experiencing wonder leaves audiences in an effectively exposed state as

---

39   For a similar discussion read Laurier, Eric. 2004. 'The Spectacular Showing: Houdini and the Wonder of Ethnomethodology', *Human Studies*, 27: 385–387. https://doi.org/10.1007/s10746-004-3341-5.

40   However, this is hardly the only way magicians make sense of knowledge of methods. Knowledge of 'how it was done' has be said to enhance the effects of (at least some) tricks. See Kestenbaum, David. 2017, June 30. 'The Magic Show—Act Two', *The American Life*. https://www.thisamericanlife.org/619/the-magic-show/act-two-31

41   Hugard, Jean and Braué, Frederick. 2015. *The Royal Road to Card Magic* (Video Edition). London: Foulsham: 10.

42   Shalmiyev, Rich. 2020, June 21. *Presentation in the 'Bridging the Impossible: Science of Magic, Wellbeing and Happiness' Workshop*.

43   Of course, in some settings (such as stage magic), performers staring into darkened auditoriums might have limited possibilities for looking at or gauging audiences.

they struggle to make sense of what they witnessed. As magicians such as Suzanne have advocated, at the culmination of a trick, audiences let their emotional guard down. To respect others, it is better therefore for magicians to look away for a beat, and then re-establish a connection after the audience has had a moment to recompose themselves.[44]

Acting in relation to the needs of others is also not straightforward because of alternative conception of the end goals of magic. As noted above, Jon Allen advocates a relaxed attitude when things go wrong because mess-ups provide a basis for developing a personal connection with audiences. And, for Allen, connection is the point. In contrast, Ortiz has called for minimizing regard for such moments in order to get on with producing strong effects:

> When something goes wrong in a performance, your job is to make the audience forget it as quickly as possible. Whining and self-indulgently dwelling on the matter will only impress the screw-up more strongly on their memory. If, instead, you treat the matter as of little importance, they will too.[45]

Such orientations offer highly contrasting ways to think about the place of vulnerability and mutuality.

Another source of trouble in caring relates to audience feedback. While magicians might be motivated to act in relation to the audience's needs, previous chapters outlined many of the reasons that experienced conjurors have identified for why this can be challenging: settings might not easily allow for anything but coarse and undependable forms of feedback (for instance, clapping); audiences can be too polite to voice criticisms directly or not be bothered enough to raise them at all; disapproving remarks can be discounted by magicians because of their pre-existing beliefs; and so on.[46] In certain respects, it is the very interpersonal considerations leading audiences to go along with someone playing the conjuror role that cast doubt on the wisdom of taking audiences' overt behavior as a reliable guide to their inner feelings. Yet, without a way to gauge participants' experiences, it is difficult for

---

44 Comments from Suzanne in Regal, David. 2019. *Interpreting Magic*. Blue Bike Productions: 424–425.
45 Ortiz, Darwin. 1994. *Strong Magic*. Washington, DC: Kaufman & Co.: 432.
46 For a discussion of many such considerations, see Brown, D. 2003. *Absolute Magic* (Second edition). London: H&R Magic Books.

magicians to be receptive to them. Today, whilst social media enables refined techniques for soliciting feedback that are not conditioned by face-to-face interactional considerations, their trustworthiness and representativeness remain open to doubt too.[47]

In my personal experience, cultivating conditions that enable meaningful feedback can be demanding. For instance, the first venue where I put on paid-for shows regularly solicited comments from audience members through a post-event questionnaire.[48] Such comments were invariably brief and positive. Although they might have bolstered my confidence as a fledgling magician, I felt I could not invest too much weight in such responses as a guide to audiences' experiences because of concerns about their readiness to criticize (see pages 69–72). Similarly, in my own experiences with others watching magic (or going to the theater, a yoga class, a restaurant, etc.), how audiences act during an event (for instance, engaged) can be markedly different from how they recount their experiences afterwards (for instance, bored). I might have even engaged in this kind of duplicitous behavior from time to time! When feedback is unreliable, it cannot serve performers to think or act in relation to others' needs.

As with other activities, caring in the case of magic is not only realized by individuals coming together. It is also constituted through an assemblage of objects: coins, handkerchiefs, chairs, tables, boxes, lighting and much besides. How care extends to such items is another matter for consideration. Many scholars working with the notion of care have sought to question commonplace tendencies to relegate the material world into a set of mere objects. In relation to promoting ecologically sound ways of living, Maria Puig de la Bellacasa argued that environmental agendas need to depart from treating soil as a productive resource that can be used according to human whims. Instead, soil should be respected as a living world with its own ecology. In this way, calls to care aim to promote considered forms of attention.[49]

---

47 Owen, Anthony. 2019, April 15. *The Insider*. https://www.vanishingincmagic.com/blog/the-insider-anthony-owen
48 All the proceeds from all the shows I have put on have been donated to charities, because that is a manifestation of caring too.
49 See Puig de la Bellacasa, M. 2017. *Matters of Care*. Minneapolis, MN: University of Minnesota Press.

In contrast, in relation to their ultimate ends, conjurors of modern magic often expend a great deal of labor to achieve the opposite orientation to the material world. As explored in Chapter 4, getting audiences to take coins, rings, ropes and other props as ordinary, off-the-rack, uninteresting and so on is often highly desired.[50] One objective of enabling you to inspect a coin, box or rope is for that object to be rendered into a genuine but still mere object: that is, a thing not worthy of much need for further scrutiny, let alone looking after. Instead, it is given, stable and transparently understood.[51]

Such attempts to render the items of magic into mere objects, though, are not without their recognized troubles as well. A danger is that audiences might act on this basis. Clients that pay for the services of magicians, for instance, might expect to keep a signed playing card, Rubik's Cube, or coin as a souvenir. If these are, in fact, specially designed props, the show might end up costing the magician.[52] In addition, rendering props into mere things is problematic because, as noted above, conjurors often advocate using objects with symbolic resonance to make the magic meaningful.[53] As a result of these kinds of competing considerations, the place of caring for the objects in magic is a delicate matter.

Care Through Promoting Discussion

In line with the overall approach in *Performing Deception* of conceiving of magic as an interplay of co-existing but contrasting considerations, the previous sections proposed some of the ways control and care get entangled. In general, to imagine the conjuror — as carer suggests the

---

50 As elaborated in Hopkins, Charles. 1978. *Outs, Precautions and Challenges for Ambitious Card Workers*. Calgary: Micky Hades: 51; and Smith, Wally. 2016. 'Revelations and Concealments in Conjuring', Presentation at *Revelations* Workshop (Vadstena) 8 April.

51 Alongside this orientation, other have used historical or personal objects as a basis for deception. See Landman, Todd. 2020. 'Making it Real'. In: *The Magiculum II*, T. Landman (Ed.). [n.p.]: Todd Landman: 48.

52 As conveyed by Allen, Jon. 2019, June 19. *Day of Magic Presentation*. Leamington Spa.

53 Another issue in orientating to the materials of magic as mere things is the way material apparatus can tune performers. As noted in Chapter 4, prominent figures have warned against rehearsing in front of a mirror because it can condition unintended, unappreciated and, ultimately, undesired ways of acting.

importance of appreciating how conjurors and audiences are dependent on one another, being receptive to audiences' experiences, and attending to any troubles experienced for the lessons they might hold. Within the dynamics of deception that constitute magic as an activity, caring for your audience can serve as a means of fooling them, and fooling them can be an expression of regard—and even, some contend, love.[54] Deception itself, though, is typically achieved through efforts of control.

To further appreciate how care and control can get entangled, the remainder of this chapter turns to strategies whereby I, as an individual novice, sought to integrate control and care in my routines. These were overt performance settings in which I played the role of an 'academic magician'; that is to say, a conjuror that sought to use magic to raise questions about the human condition,[55] in large part through referring to scholarly ideas and concepts. Instead of seeking to re-enchant the world through demonstrating wondrous feats, the overall intention that emerged was to foster an appreciation of the mundane, 'seen but unnoticed',[56] and tacit ways we act together.

Let me elaborate how by starting with the overall design. As developed in Chapter 3, the basic focus group model adopted for my recorded sessions and public shows was intended to take engagement with audiences beyond the typical affective responses that follow acts of magic (for instance, displays of surprise, curiosity, incredulity). Participants were asked to reflect on our interactions as they unfolded and such reflections helped constitute those very interactions. In part, this was done by posing questions to the audience after the culmination of each effect. Those present then offered reflections for group discussion. In both my recorded sessions and public shows, this overall design served as a central basis for engaging with others in the moment and thus being responsive to what was arising for them. The conversation also served as the basis for subsequently revising the content and delivery of the sessions. Through doing so I sought to tailor the magic around others.

---

54 See comments from R. Paul Wilson comments in Regal, David. 2019. *Interpreting Magic*. Blue Bike Productions: 544.

55 For one elaboration of this notion, see Landman, T. 2018. 'Academic Magic: Performance and the Communication of Fundamental Ideas', *Journal of Performance Magic*, 5(1). https://doi.org/10.5920/jpm.2018.02

56 To adopt an expression from Garfinkel, Harold. 1984. *Studies in Ethnomethodology*. Cambridge: Polity.

However, as discussed in Chapter 3, while our interactions generated group dialogue, treating the resulting conversations as unadulterated expressions of inner thoughts and feelings would be problematic. This is so, in part, because of the points made in the previous section about the unreliability of feedback. During the delivery of these shows, questions also sprang in my mind from more generic concerns about the pervasiveness of 'impression management' in exchanges. Fields such as social psychology and sociology have long suggested that interpersonal communications are pervaded by defensive mindsets in which individuals attempt to avoid themselves or others being threatened.[57] This can lead to covert attributions of motives, the orientation to one's thinking as obvious and correct, the use of face-saving expressions and so on.

Another source of caution related to how the discussions in my sessions were managed. One claimed advantage of focus groups as a method of research is that they enable those moderating the dialogue 'both to direct the conversation towards topics that you want to investigate and to follow new ideas as they arise'.[58] How moderators reconcile the desire to steer and be steered, though, is a conundrum that has to be worked out in practice. Frequently. The basic need to reconcile these desires undercuts any notion that a focus group format simply enables participants to express themselves in their own terms.

In short, the magic sessions involved a doing together, but this took place in highly managed and mediated interactions in which questions can be asked about how deception, truth and caring comingled. The remaining sections of this chapter turn to such questions through examining how the intent to be responsive related to how attention was directed, how manipulation was achieved and how interactional troubles still emerged.

Discussing Attention and Challenge

Let me start by reviewing additional aspects of the focus group format, through which I tried to solicit and be responsive to the audiences'

---

57 See, for instance, Argyris, C. 2006. *Reasons and Rationalizations*. Oxford: Oxford University Press. https://doi.org/10.1093/acprof:oso/9780199268078.001.0001.
58 Morgan, D. 1998. *Focus Groups as Qualitative Research*. London: Sage: 58.

experiences. As with magic in general, the focus group-type dialogues in my sessions were acts of directing regard. As mentioned already, one of the ways I directed attention was to ask participants to consider how magic was something we achieved together through mundane actions and inactions that might well be 'seen but unnoticed'.[59] Through making what was taking place between us into a topic for joint consideration,[60] I sought to cultivate the possibility to be moved by and to respond to others.[61]

Take matters of attention and challenge. Before I started performing magic, I had worried about how closely audiences would attend to my actions and how often they would intervene to disrupt them. As I soon concluded, though, attention and challenge were not forms of audience behavior that I had to minimize in order to ensure effects could be pulled off. Instead, I needed both. Audiences had to follow along closely enough to be able to be amazed at the final outcome. Similarly, audience interventions during magic effects heightened the sense of their improbability. And yet, if pursued too robustly, attention and challenge would have made the effects impossible to pull off.[62]

Having derived these observations throughout the initial sessions I conducted in 2018, I began asking groups in subsequent sessions to offer accounts of how they were attending and challenging. As previously noted in Chapter 3, participants often accounted for their lack of interventions by contending that they were deliberately working to contribute toward the success of the effects. After hearing such justifications, in subsequent sessions I asked participants to reflect on how they were (and were not) attending to and challenging my actions through referencing the prior justifications given by other participants in earlier sessions.[63] Thus, instead of simply seeking to gather accounts

---

59 To adopt an expression from Garfinkel, Harold. 1984. *Studies in Ethnomethodology*. Cambridge: Polity.
60 Zimmerman, Don H. and Pollner, Melvin. 1971. 'The Everyday World as a Phenomenon'. In: *Understanding Everyday Life*, J.D. Douglas (Ed.). London: Routledge & Kegan Paul: 80–103.
61 Hendriks, Ruud. 2012. 'Tackling Indifference—Clowning, Dementia, and the Articulation of a Sensitive Body', *Medical Anthropology*, 31(6): 459–476. https://doi.org/10.1080/01459740.2012.674991
62 A tension explored in-depth in Hopkins, Charles. 1978. *Outs, Precautions and Challenges for Ambitious Card Workers*. Calgary: Micky Hades.
63 In promoting this kind of situated telling, I was able to de-individualize my questioning of participants conduct.

from participants about their behavior there and then, I introduced my observations and reflections derived from previous experiences. This was done to promote group inquiry.

Discussing Eye Contact

Take another example. With the spread of Covid-19 in the spring of 2020, like many magicians, I pivoted toward offering Zoom-based online shows. Sixteen sessions were held through the Ashburton Arts Centre and the Exeter Phoenix arthouse up until February 2021.

Such technologically mediated forms of delivery raise many questions about how those present meet one another. For instance, eye contact is vital for establishing rapport and trust in many inter-personal relations. Platforms such as Zoom both enable (the appearance of) eye contact between magician and individual participants, and frustrate contact. They enable it in the manner each participant can see the magician directly in front of them, no matter the audience size. They frustrate eye contact because magicians are drawn to look down at the people on the computer screen so as to view their reactions, rather than up into the camera lens so as to be seen to be looking at audiences (from the latter's perspective). In response, some conjurors have proposed various solutions to establish the pretense of eye contact. This pretense is secured through compelling magicians to look into the camera lens rather than at the computer screen. These solutions include shielding the computer screen, positioning the participant image window as near to the camera as possible, and reconsidering whether this delivery platform is appropriate.[64]

In my online shows, I adopted a different orientation. Rather than seeking a solution for how to look, I sought to make the conditions for eye contact into a topic for discussion. The impetus for doing so stemmed from the first time I practiced for others on Zoom. In this session, I used a camera that inclined down onto my card table for participants so that audiences could closely scrutinize my card handling. A friend suggested the camera view needed to change. One reason she cited for doing so

---

64 Houstoun, Will and Thompson, Steve. 2020, June 7. 'Are You a Prioritisationalist?', *Video Chat Magic*. See https://videochatmagic.substack.com/p/are-you-a-prioritisationalist

was that seeing my eyes enabled her to imagine I was addressing her directly. What proved particularly interesting for me was what she then went on to say: that by seeing my eyes looking at her she could experience a one-to-one connection with me even as she recognized full well that I was rarely directly attending to her image. In other words, when I showed my face and looked into the camera, she felt an affective bond despite the conscious recognition that we were almost assuredly *not* making 'eye-to-eye' contact.

Based on this feedback, I made the constitution of inter-personal connection into a topic of group conversation. I typically did so in this way: at the end of one effect, I asked participants whether they were confident that they were seeing all that they needed to see in an online delivery format. Since virtual shows do not allow for the same kind of scrutiny as face-to-face interaction, I wanted to hear whether participants thought online delivery offers any additional possibilities for magicians to deceive. However, I then asked them if they were seeing *more* than what was taking place. I did this by relaying the participant's comments in the previous paragraph and then illustrating how the scene appeared to them as I varied my gaze between the recording camera and my computer screen. In making the conditions for eye contact into a topic of group conversation, I sought to open up a space for discussing participants' ongoing experiences regarding how we as a group made the activity of magic happen together online.[65]

In short, a kind of 'meta-magic' was sought by making the basis for our interactions into topics for discussion. Audiences were asked to consciously step back from and monitor their conduct and beliefs. The lines of questioning for doing so were developed through iterative cycles of action, consideration, preparation and revised action that sought to devise interactions responsive to others.[66]

---

65   In support of promoting wider reflection about such topics, in late 2020 I produced a website featuring recorded routine clips and accompanying text entitled *The Magic of Social Life*. See https://brianrappert.net/the-magic-of-social-life

66   As part of my experimentations in directing attention, for three online shows conducted through the Ashburton Arts Centre in 2020 entitled 'Magic: Who Cares?', care in magic was taken as the principal substantive focus.

Care Through Manipulation

Connection, attention and challenge were not only topics for reflection in our dialogue, though. Regard for them also served as a resource for accomplishing trickery.

For instance, dialogue was used to condition subsequent interactions. To elaborate, with experience I began to place discussions about the limited extent of audience challenge before effects that were enhanced by animated physical participation by the audience. By making challenge into a topic for group reflection at one point in time, I sought to encourage challenging forms of behavior at the next point. My inspiration for doing so was an experience in the 13th recorded face-to-face sessions discussed in Chapter 3. As part of this event, I described the limited extent of challenge I had experienced in prior sessions. During the next effect, the person I was working with went on to select a card other than the ones spread out close to her, thereby disturbing the sequencing of cards that underpinned the methods for the effect. Some minutes later she justified her actions to the group by citing the previous discussion about the limited challenge as extending an invitation for her to do so:

Excerpt 7.2—Session 13

No	Direct transcript
1	P4: But can I ask you? Because I feel really terrible, because I sort of ruined your last trick. And, and when we came here we sort of we were talking that we had a contract, almost that we were going to watch you and we were going to be amazed by your tricks and we were not going to destroy them. But then you told this story about people who <u>had</u> challenged you
2	BR: Yes, yes
3	P4: Yes, and then this it sort of opened up the
4	P4: [possibility
5	BR: [Yes, yes
6	P4: myself for me to. Is that also part of your project? Have you seen that before?

7. Control and Care 165

Up until that stage, the kind of openness P4 spoke to in Lines 3, 4 and 6 was not part of my project. Based on her comments, though, I started deliberately to position the group discussion of challenge so as to promote physical interventions by the audiences: interventions such as mixing up cards in an animated and thorough fashion. The purpose of doing so was to enhance the experience for audiences.

Another dimension of how the discussion of attention and challenge served as a manipulative resource was the way both functioned as forms of misdirection. Toward the end of the 30 recorded sessions, I positioned the discussion of attention and challenge before effects with methods that were resistant to being foiled by close attention or audience disruption. However, in my patter and mannerisms, I portrayed the effects as reliant on the precise control of cards and the choices of audiences. Pre-framing effects in this way also encouraged our subsequent group discussion to return to themes about attention and challenge.

Still another manner in which regard to attention and challenge served as a resource was by using my verbal patter about attention and challenge to provide a cover for the control of cards. For instance, in one face-to-face session the following interactions took place:

Excerpt 7.3—Session 14

No	Direct transcript	Non-verbal actions
1	P1: other than doing what we are told, I think we are pretty (0.5) passive (3.0) players in the (.) magic.	BR picks up deck
2	BR: Hum, hum. One of the things I am interested in is attention and the way attention kind of gets negotiated in these sort of settings. So did, did you bring up attention before, right?	BR spreads the deck
3	P1: Hmm.	
4	BR: Okay, so, I mean, I could be going through the deck like this or something like that and, umm, you might be at times really focused, okay. Other times maybe looking around,	BR leans in BR leans back and completes a card sleight

No	Direct transcript	Non-verbal actions
5	BR: [right?	
6	P1: [Hmm.	
7	BR: So it has been interesting for me cos this is the first time I have done <u>these</u> tricks in this way. But then we had, I did have people around before and did a different kind of routine ((inaudible)) And, you know, there are like two extremes. One extreme like was this guy (2.0) on his mobile phone for most of the evening going like this.	*BR pretends to be using a mobile phone slightly under the table*
8	P1 and AU: ((laughter))	
9	P1: That's really edifying.	
10	((laughter, side conversation))	
11	BR: And then the other extreme was, was when uhm, uhm I was doing these and someone said, umh, she said, oh Brian, I'm (.) watching you and the cards and I am watching. And then she kind of leans in like this. Okay, she did not watch the whole night, obviously, but you know for quite a bit of it she was just like this.	*BR leans forwards* *BR leans back* *BR leans forward and peers downward*
12	P2: I'm watching you pretty hard actually.	
13	BR: Okay=	
14	P3: =He has not seen the movie yet, so he has not noticed yet ((laughter))	
15	BR: But, it is not, you are not totally watching me. I mean you ar- are talking	
16	BR: [to	
17	P1: [Yeah	
18	BR: P1 or P2.	
19	P2: I'm watching you pretty hard actually.	
20	P1, P3 and BR: ((laughter))	

No	Direct transcript	Non-verbal actions
21	BR: You are watching me pretty hard. I can feel the heat coming out.	BR *shakes hands*
22	P1, P3 and BR: ((laughter))	
23	P1: That's mildly threatening ((P2))	
24	((side conversation))	
25	P1: ((P2)) has slight paranoid tendencies.	
26	P2: I am always interested in looking for the angles on things	
27	BR: Hmm	
28	P2: And I really dislike being a mug.	
29	P1 and P3: ((laughter))	
30	P2: And I have a slightly flawed relationship to this sort of thing. Because it plays into that a little bit.	
31	P1, P3 and BR: ((laughter))	

In this exchange, under the guise of spreading the deck out in my hands and inviting participants to witness my doing so, I was able to spot a card needed for the next effect (during Line 4). Then I was able to control it to the desired position through a hand movement during my subsequent gross bodily movement of leaning back from the table (Line 8). My verbal remarks pointing toward what was taking place at that moment (Line 4) and the accounts of previous participants' attention (Lines 7, 11) functioned to preoccupy those present and thereby curtail their ability to later reconstruct what had taken place. It was through such actions I was able to achieve inexplicable feats for audiences.

Caring About Troubles

As outlined in the previous section and Chapter 3, a consistent part of the rationale for my shows was to foster dialogue with participants regarding their experiences as we interacted together under the label of doing magic.

As some have advocated, 'understanding requires an openness to experience, a willingness to engage in a dialogue with that which challenges our self-understanding. To be in a dialogue requires that we listen to the other and simultaneously risk confusion and uncertainty both about ourselves and about the other person we seek to understand'.[67] Through the overall design, I sought to achieve a format that provided the basis for exchange that made my understandings and uncertainties into topics for dialogue. Through doing so I sought to fashion shows in accordance with the responses of others.

As suggested previously, care as a relational practice of attention requires a willingness to acknowledge ethical troubles. I turn now to some such troubles.

For a start, to be sure, the shows did not realize an idealized form of freely open dialogue. The philosopher Martin Buber, for one, contended that authentically being with others requires the absence of deception.[68] In stark contrast, the routines were constituted through deception.

Likewise, while I sought to devise effects and exchanges that would be responsive to my emerging understanding of others' concerns, it was me who steered this development. I regularly realized an asymmetrical influence over what was discussed, who spoke and for how long. As such, the exercise of care and control came bundled together. As indicated by the quote at the start of this chapter from my first public show, not everyone found the type of coordination taking place appropriate.

Other interactional troubles can be identified as well. For instance, despite my initial 30 recorded sessions examined in Chapter 2 taking place (largely) between known acquaintances as a form of research-entertainment, they were not without ethical knots, binds and discomforts. In an early session, one person became agitated to the point of repeatedly getting up from the table because the effects reminded him of childhood experiences of being humiliated by magicians. His action was verbally sanctioned by his partner. Various participants offered apologies during and after the sessions for behavior which they

---

67 Schwandt, TA. 1999. 'On Understanding Understanding', *Qualitative Inquiry*, 5(4): 458. https://doi.org/10.1177/107780049900500401.
68 Buber, Martin. 2018. *I and Thou*. London: Bloomsbury Academic.

thought fell short of what was expected of them. Lines 28–30 in Excerpt 7.3 regarding being 'mugged' also hint at some of the fraught ethical and affective dimensions of trickery. This includes the potential for individuals to feel defensive, duped, demeaned and so on. Along these lines, in the last of my 30 recorded sessions, an effect involved one of the participants orchestrating the other participants to eliminate all the cards from a face-down deck one-by-one until only a single card remained. I proposed that the remaining card would be a card previously signed by a participant. The person coordinating the selection process did so with an extraordinary degree of meticulousness—the selection process lasted over five minutes. When it became apparent this process was just an extraneous set-up for a follow-on effect I undertook, she commented that she felt 'cheated' because the selection was 'all for nothing'. While these comments were jokingly delivered, I certainly felt awkward at the time and openly commented so.

Additional kinds of ethical troubles emerged in the recorded sessions because they were not only magic displays. Instead, I was undertaking formal research. As such, the researcher-participant relationship became entangled with the magician-audience one. In this regard, consider a basic distinction in how we attend to one another. The philosopher Nel Noddings juxtaposed *projection* and *reception*. *Projection* involves efforts to analyze and establish what another is experiencing. As such, it entails a form of objectification. Such objectification is routinely built into social research. *Receiving* the other, in contrast, requires a motivational shift. It calls for becoming engrossed with the other to attempt to feel for and become sensitive to their wants and needs, even as it is recognized that it is not possible to straightforwardly access their experiences. Noddings argued that reception is not about making another person into an object, because reception is not driven by a desire to make claims to knowledge. Instead, it entails an openness to be transformed by others.[69] In making the case for the importance of reception in caring relations, Noddings did not seek to exclude other forms of attending. Analytical forms of projection to derive knowledge still have an appropriate place. 'What

---

69 To offer a different language, the distinction between projection and reception brings to the fore the question of whether we treat our interactions with others as ends in themselves, or as a means to some agenda.

seems to be crucial' she argued 'is that we retain the ability to move back and forth and to invest the appropriate mode'.[70]

Concerning the recorded sessions, initially I was highly indebted to forms of attending based on projection. That is to say, I engaged audiences in order to analyze their experiences. I did so, not least, to formulate research findings such as those given in *Performing Deception*. I pressed participants to give accounts of their experiences that could serve as data and experienced the inability to gather such data at the time as a failure on my part. Adopting this orientation risked reducing mutually responsive interactions into an effort to extract data from research subjects. With my gradual recognition of how I was tied to relations of projection, later I refrained from recording some pre-arranged magic sessions so as not to get wedded to projection.[71]

While attempts to reconcile projection and reception caused tension felt throughout my recorded sessions, additional aspects of my relationship with others on matters of care underwent distinct development. For one, when I first began doing magic for others, I invested a great deal of effort in ensuring that the intended outcomes (for instance, card identification) were obtained. That was my working sense of what it meant to think and act concerning others. As I developed, however, the goal of 'getting it right' gradually gave way to the goal of engaging with and responding to others. Such relations could be accomplished even if the effects failed by some conventional performance measure. 'Botched tricks' brought their opportunities for reflection relating to the matters that were of interest to me and others.

Noteworthy too, in many instances where something had 'gone wrong', participants frequently blamed themselves for not having acted correctly.[72] In such ways, those being 'cared for' overtly contributed to creating a caring environment. Eventually, I incorporated effects into my routines in which I could not fully control the outcome. When the

---

70 Noddings, N. 2013. *Caring* (Second Edition). London: University of California Press: 35. https://doi.org/10.1525/9780520957343.
71 And in a number of instances, I refrained from performing magic so as not to treat friendships and other relations as opportunities for practice and training.
72 Failure also brought opportunities for manipulation. Even for effects where I should have been able to control the outcome, when things went awry I sometimes went on to prompt group reflection and their responses to the blunder, as if I had planned all along that the effects would not work out.

intended outcome was achieved, the magic on display was arguably very strong. When the outcome was not achieved, I used the failure to prompt group reflection on how participants responded to the failure (for instance, offering apologies for their behavior, looking away, changing the topic[73]). Promoting group reflection in this way provided a basis for considering magic as a joint accomplishment.

Likewise, to understand the demands participants felt when playing the role of an audience member, I came to ask them as part of my questioning about the vulnerabilities they experienced and the kinds of emotional labor they undertook in playing the 'audience member' role.[74]

In such ways, I was able to make vulnerability into a topic of conversation. Other kinds of vulnerability proved less adaptable. My most emotionally charged moments came, not from when the effects went awry, but rather when I felt participants attentively disengaged while I was trying to engage them. Side conversations, stares into the distance and scrolling on mobile phones were some examples of what I took to be disengagement. In other words, the strongest affective charge was associated with conditions of responsiveness rather than the content of specific actions. When participants were no longer concerned with undertaking the kind of work needed to sustain and coordinate our relations, our time together could no longer be understood as interaction. This is something I cared about maintaining. As I have come to understand my reactions, they stand as further evidence for the mutual dependencies between conjurors and audiences.

In Close

Against the frequently aired contentions that magic requires conjurors to remain in control, this chapter has asked how its undertaking can be approached through the notion of care. As initially argued, while 'care' is not a word often heard in gatherings of conjurors, they often ask how they should act in relation to their audiences.

In line with prevalent academic theorizing, care has been understood as a willingness to think and act in relation to the needs of others. More

---

73   For a discussion of this point see *The Magic of Social Life —Vulnerability* at https://brianrappert.net/the-magic-of-social-life/vulnerability
74   See *ibid*.

than this, to care requires posing questions about how caring takes place, what it means to be responsive to others, how the cared-for contribute to caring, the power asymmetries in who defines care as well as varied other issues. As contended, conjurors frequently attend to such matters in thinking about how they ought to go about manipulating their audiences.

A central preoccupation of this chapter has been to characterize how control and care are bundled together in complex ways in the relations between magicians and audiences. Those ways underscore a theme that has run throughout *Performing Deception*: the importance of treating the performance of magic as a form of reciprocal action.

More than just bringing together professional and academic arguments related to care and control, this chapter has examined the evolving manner whereby I sought to bring control and care together in my performances. It has mapped out how I sought to develop my responsiveness to audiences across my initial 30 recorded research sessions as well as my face-to-face and online public shows. To become more responsive, I fostered certain kinds of attentiveness by:

- soliciting feedback on the performances by directly asking participants about their thoughts and feelings. With this feedback, I was able to modify my subsequent performances;
- offering accounts of the actions of past audience members to encourage inquiry into the current actions;[75]
- marshalling questions and observations so as to enable deception and to shape participants' actions.

Through these strategies, I sought a form of magic that was self-referential on two levels. Firstly, like with many other forms of entertainment magic, I portrayed the tricks as tricks. In other words, rather than being down to genuine extraordinary powers, I openly acknowledged the operation of secreted methods at work in the accomplishments of

---

75 The first two of these were in line with Chris Argyris' notions of single- and double-loop learning; see Argyris, C. 1999. *On Organizational Learning*. Cambridge: Blackwell.

effects.[76] Secondly, though, I also sought to promote reflection on how magic was accomplished together by magicians and audiences. This consideration about what was taking place there and then between us, in turn, served as a basis for accomplishing our interactions.

---

76 That is, after my initial round of recorded sessions which were themed around embodiment.

8. Learning and Unlearning

Performing Deception began by taking an activity as its object of attention, namely entertainment magic. Successive chapters have detailed how the crafts of conjuring are learnt through recounting the experiences, abstractions, reflections and experimentations of a novice, as well as of seasoned practitioners.

In this concluding chapter, I want to continue in the spirit of treating learning as a process of iterative development by first returning to the starting topic for this book—now informed by the previous chapters.

What, then, is entertainment magic?

In responding, it is important to first acknowledge that what counts as an appropriate answer depends on the reasons for posing the question. In this regard, let me begin by noting some prominent scholarly depictions of magic and the wider intellectual projects associated with them.

The philosopher Jason Leddington has sought to establish what makes magic a distinctive and unique aesthetic experience.[1] For him, entertainment magic is first and foremost concerned with displays of the impossible.[2] While conjuring might incorporate comedic or theatrical moments, these features are not the marks of magic. Instead, what distinguishes magic from other activities is the conjuror's intention to create illusions of the impossible. Furthermore, this sense of impossibility is not the make-believe associated with reading fictional novels or watching Hollywood blockbusters. Instead, 'it is essential to a

---

1 Leddington, Jason. 2016. 'The Experience of Magic', *The Journal of Aesthetics and Art Criticism*, 74(3): 253–264 and Leddington, Jason. 2020, May 28. 'Savouring the Impossible', *Aesthetics Research Centre Online Seminar*. http://aesthetics-research.org/archive/2020/leddington/.

2 See, as well, Coppa, Francesca, Hass, Lawrence and Peck, James (Eds). 2008. *Performing Magic on the Western Stage*. London: Palgrave MacMillan: 8. https://doi.org/10.1057/9780230617124.

magic performance that *impossible events actually appear to happen.*'³ The result is a cognitive bind: audiences know that what is happening is impossible, the magician presents it as impossible, and yet it appears to be taking place nonetheless. The combination of those beliefs and displays creates an oscillation between confusion and curiosity.⁴

Integral to achieving a sense of the impossible for Leddington is the requirement that magicians cancel out every explanation that audiences might harbor. For David Copperfield's flying through the air to be magical, for instance, the performance must negate each of the premises audiences hold about how his movements could be achieved. The belief that he is suspended from wires, for instance, needs to be negated by Copperfield moving through alternatively aligned metal hoops. Without such cancellations, the performance might be regarded by audiences as impressive, but it should not be labelled as magical.⁵

In offering these arguments, Leddington provides a variety of distinctions for marking out what is specific to magic. The identification of distinctions means it can be contrasted with the essential qualities of other aesthetic experiences.

Other scholars have taken alternative aims. In *Magic's Reason*, anthropologist Graham M. Jones takes as his concern how entertainment magic has been varyingly understood within European traditions of thought.⁶ As he argues, over time its meaning has been entangled with what counts as occult magic. Whilst entertainment conjurors have sometimes sought to tap into the mystique of the occult, the prevailing

---

3   Leddington, Jason. 2016. 'The Experience of Magic', *The Journal of Aesthetics and Art Criticism*, 74(3): 255. https://doi.org/10.1111/jaac.12290. (Emphasis in the original).
4   An oscillation that can be used to bring into effect wonder and thrill, or wonder and unease. See Taylor, N. 2018. 'Magic and Broken Knowledge', *Journal of Performance Magic*, 5(1). https://doi.org/10.5920/jpm.2018.03.
5   In my experience, realizing the impossible sets a high bar for effects —one that seems to rule out a great deal of activity labelled 'conjuring'. By proscribing that magicians must cancel out each of the likely explanations audiences harbor, this definition would exclude the vast majority of effects I have encountered in books, DVDs, conventions, etc. devised by leading magicians. Whilst effects often counter a limited number of probable explanations, few of them systematically ensure each and every explanation is cast into doubt. Also, audiences might regard some feats of magic as impossible, such as turning one object into another. Especially in the case of card magic in which laws of physics are rarely at stake, however, 'improbable', 'adroit' or 'inexplicable' seem more apt labels for the activities taking place.
6   Jones, Graham M. 2018. *Magic's Reason*. Chicago: Chicago University Press. https://doi.org/10.7208/chicago/9780226518718.001.0001.

tendency going back to at least the mid-19th century has been to oppose secular, entertainment forms of magic with so-called primitive ones. Through doing so, entertainment magicians have aligned themselves with notions of the modern and the rational. Jones' task in *Magic's Reason* is not only to recount a history of magic within the development of modernity, but to relate this to the development of social anthropology. Within the latter field, magic has been a central topic of study. However, as Jones contends, attention has been cast overwhelmingly toward occult forms. With this selective preoccupation, social anthropologists have portrayed belief in magic as relevant to the irrational and primitive, while they have sought to cast themselves as rational and modern through their efforts to explain other cultures.

In forwarding his argument, Jones uses the question of what counts as magic to inform the understanding of high-level concepts: modernity, rationality, ritual, culture and so on. In this pursuit, he has not been alone. Chris Goto-Jones took up the relation between modernity and magic by examining the tension-ridden manners in which the oriental magic of China, Japan and India was embraced, diminished and appropriated within Anglo traditions.[7]

In contrast to these projects, *Performing Deception* has sought to understand some of the practical forms of reasoning and skills associated with conjuring. In particular, I have examined some of the ways it is taught and learnt through instances of demonstrating, instructing, performing and the like. As noted, demonstrating, instructing, performing and the like rely on routine sense-making processes, even as magic underscores how sense-making is fallible. In this concluding chapter, I want to return to some of the premises for this study as well as offer conclusions that follow from it.

---

7   See, as well, Goto-Jones, Chris. 2016. *Conjuring Asia: Magic, Orientalism, and the Making of the Modern World.* Cambridge University Press. https://doi.org/10.1017/cbo9781139924573. For an analysis of how magicians have figured as archetypal figures in cultural imaginations, see Granrose, John. 2021. *The Archetype of the Magician.* Agger: Eye Corner Press.

Joint Wonders

As part of the agenda pursued here, reasoning and skill have been regarded as practical doings. I have sought to examine how they are realized in and through situated actions. This orientation has shaped how notions such as culture, expertise and naturalness are conceived. Much of my effort in the previous chapters has been dedicated to elaborating how the realization of culture, expertise, naturalness and so on takes place through verbal communication, deliberate gestures, body orientations, object placement, directed gazing and so on.

Central to this analysis has been the contention that magic is not something done only by magicians. Instead, those acting in relation to roles such as 'audience member' and 'magician' realize a sense of their identity through each other. In short, it is a joint activity, albeit one that typically involves stark asymmetries in knowledge and action. In particular, my focus has been on forms of group encounters. I have sought to understand these occasions as entailing mutual dependencies wherein each person's experiences is dependent on the others present, as well as on the evolving group situation. Accordingly, proficiency is realized through relations with others, rather than being an attribute of those billed as 'performers'. Even when conjurors practice alone, magic is not well understood as an insulated activity.

Aligned with this general orientation, a recurring theme throughout this book has been how those partaking in magic can know what others are thinking, wishing or feeling. As a performance art, conjurors attempt to put themselves in the place of their co-present, virtually present or imaginary audiences. Doing so is problematic, not only because of general questions that might be asked of how any person can know another, but because magicians engage in actions designed to create an experiential divide between themselves and others. *Performing Deception* has taken as a central concern the reasoning and skills for an activity in which audiences generally accept that deception and manipulation are afoot.

As elaborated through this 'self-other study', the work involved in trying to know another varies considerably across encounters. Chapter 2 spoke to the forms of envisioning which take place in reading written instructions for tricks. It also suggested that how appreciating what instructions can instruct involves becoming awareness of their limitations. Such limitations stem from the inability of instructions to guide decisions about how to act in relation to the unfolding expressions, positioning and other actions undertaken by audiences. Furthermore, enacting instructions places demands on readers to bring to bear standards beyond those provided by the instructions themselves. However, it is just these kinds of appreciations that are not available to novices.

In recounting my first attempts to perform face-to-face magic, Chapter 3 relayed further conundrums in trying to know others. These sessions involved a dynamic interplay between separation and connection. As in social life more generally, the potential for establishing meaningful connections relied on the starting separation between participants and I.[8] Our sense of separation was evident in the very means we sought to overcome separation in our roles as magicians, audience members, speakers, listeners and much more besides: through aligning our bodies, gazing eye-to-eye, sequencing verbal communications, etc.[9] In the case of these sessions, specific factors regarding separation and connection were relevant. The participants and I were divided through our alternative understandings of the methods for the tricks, even as we took part in a common endeavor. The suspicion that deception was enacted through our movements and words was also a topic of concern in how it created a gap between us, even as we sought to communicate through movements and words.

Through reflecting on my initial encounters with instructions and performances, as well as recounting the writings of prominent professionals, Chapters 2 and 3 elaborated how greater familiarity with performing magic engenders a sense of moving closer to and away from appreciating the experiences of others as well as one's self.

The work involved in knowing one another was also touched on in Chapter 4. As elaborated, through highly choreographed movements,

---

[8] Baxter, Leslie A. and Montgomery, Barbara M. 1996. *Relating: Dialogues and Dialectics*. London: Guilford; and Arundale, Robert. 2010. 'Constituting Face in Conversation', *Journal of Pragmatics*, 42: 2078–2105.

[9] As is the case elsewhere. For instance, see Heath, Christian. 1984. 'Talk and Recipiency: Sequential Organization in Speech and Body Movement'. In: *Structures of Social Action: Studies in Conversational Analysis*, J.M. Atkinson and J. Heritage (Eds). Cambridge: Cambridge University Press: 247–265. https://doi.org/10.1017/cbo9780511665868.017.

magicians operating in the modern style seek to render their actions natural according to cultural conventions of the day. Achieving naturality is a way to make the actions easily recognizable and intelligible, and, thus, unworthy of note, even as audiences might well harbor the suspicion that something untoward is going on. Chapter 4 also discussed how knowing another is a thoroughly materially mediated activity in which learners need to shift between ways of feeling and sensing.

Chapter 5 began by detailing contemporary contests over who can speak for audiences and who can assess the quality of magic—seasoned entertainers, lay spectators, experimental psychologists and others. These arguments provided the impetus for investigating how the reliability and fallibility of perception are made relevant within specific undertakings of magic. In also recounting the instructions as part of a masterclass, I sought to illustrate how instructors can adopt varied and shifting orientations to perception. Students of magic can be both invited to rely on their senses in a matter-of-fact way, and warned of the dangers of doing so.

Chapter 6 examined how prominent magicians have made themselves known through autobiographies that varyingly suggested that there was more going on than appeared on the surface. In forwarding more or less stable, known, definitive images of themselves, the autobiographers also forwarded images of their audiences. As a final exploration of self-other relations, Chapter 7 turned to how individuals can and should be together in acts of deception and manipulation.

Positioning Methods and Theory

Throughout these chapters, notions of self and other have been understood as formed through co-existing and conflicting features such as separation and connection. I have sought to characterize how such features interplay. Herein, multiple kinds of methods have been invoked: magicians have their methods for simulating and concealing the basis for tricks. Audiences, too, have methods for making sense of what is displayed and for detecting conjurors' methods. Moreover, I have offered a conception of magic as a kind of method for understanding ourselves and others. This is not a method for making others or even one's self transparently known. As noted, to hide and simulate, conjurors utilize many of the same kinds of physical movements and verbal justifications that signal openness and sincerity. Audiences can do much the same. Each can have qualms about the trustworthiness of others. This analysis has suggested how doubt, acceptance, suspicion and trust mix and meld through examining how magicians[10] and audiences get entangled with each other. As a result, to characterize magic as a method is to signal its fraught potential to foster insights into ways of doing and being.

Also, throughout the chapters, examining forms of reasoning and skill has not been conceived of as a straightforward task of applying a particular scholarly theory. For instance, ethnomethodologist Eric Livingston has contrasted different types of sociologies: those of the *hidden social order* and *witnessable social order*. The former seeks to get underneath what is visibly taking place by employing methods and theories that can explain the root societal forces that shape action. In contrast, sociologies of the witnessable social order seek to describe how the orderliness of life is sustained through a detailed analysis of what is readily observable in the here and now. This is done without recourse to the theoretical frameworks and methods commonplace in sociologies of the hidden order that seek to explain one phenomenon (say, religious belief) through reference to yet other ones (say, gender).[11]

In taking the development of reasoning and skill as the prime matter of attention, I have not adopted either 'theory' or 'observable action' as an exclusive or principal framing path for inquiry. Relatedly, I have not set out an approach to inquiry based on either establishing experiments with a definitive hypothesis or describing naturally occurring social phenomena. Instead, learning magic has been treated as I experienced it: that is, as an ongoing, back-and-forth and dynamic process of relating concrete experiences, abstract concepts and theories, active experimentation as well as observations and reflections.[12]

---

10 For further commentary on how audiences can be strangers to magicians, see Tamariz, Juan. 2019. *The Magic Rainbow*. Rancho Cordova, CA: Penguin.
11 Livingston, E. 2008. *Ethnographies of Reason*. London: Routledge: 123–130. https://doi.org/10.4324/9781315580555.
12 In these broad terms, the account offered here of mixing concrete experiences, abstract concepts, active experimentation as well as observations is in line with how professional magicians recount their experiences; for instance, see Tamariz, Juan. 2019. *The Magic Rainbow*. Rancho Cordova, CA: Penguin.

This four-part breakdown of the modes of 'experiential learning',[13] while inevitably open to question for how it carves up learning, has served the purpose of drawing distinctions and relations between the undertakings entailed. In this spirit, too, proficiency in conjuring has not been conceived as simply the knack associated with controlling one's body or material objects. Instead, skill in its broadest sense has been treated as including the potential to relate experiences, abstractions, experimentation and reflections as part of emerging relations with others and the world. This capacity itself derives from previous efforts to relate experiences, abstractions, experimentation and reflections, and it conditions subsequent such efforts. I have been able to elaborate on this kind of emergent approach to skill by examining my practical undertakings over time as a learner.

In this way, rather than seeking to adopt a position somehow external to the activity of magic, I used my fledgling membership in the category of 'magicians' as a basis for understanding. This has been done even as I have sought to make what it means to be a magician or do magic into topics for inquiry.

One implication of this research design is that, rather than advancing a single theoretical framework for understanding magic, *Performing Deception* has relayed the circuitous ways abstractions can inform a sense of what is taking place in conjuring. Also, rather than treating magic as a singular (albeit perplexing) object of study, I have been interested in the tremendously varied kinds of work that achieve outcomes deemed 'magical'. As such, magic was not treated as something that exists out there in the world waiting to be discovered and inspected. Instead, what counts as conjuring is continuously made and remade through our doings—what we choose to regard as astonishing, how we behave during interactions, how we define categories and concepts to make sense of the world and so on. My unfolding doings as a learner not only shaped the sense of what I observed but shaped myself as an observer. In this way, an underlying premise and conclusion of this book is that the known and knower cannot be separated.

---

13   Kolb, D.A. 2015. *Experiential Learning* (Second Edition). Saddle River, NJ: Pearson Education.

A Heuristic Definition

Informed by my investigations as a student, I have sought to characterize magic as a *deft contrariwise performance*. I have not done so to set out a definitive, for-all-purposes, singular representation. Instead, I have offered this phrasing to cultivate sensitivities that enable us to attend to magic as a social and material accomplishment. Each chapter has sought to appreciate how notionally opposing tendencies in magic interplay and, in doing so, potentially contribute to and complement each other. How can performers learn to recognize naturality? How can they appreciate the limits of human perception through their perception? How does competitive scrutiny rely on cooperation?

These are the types of questions pursued in this study, a study that has taken paired notions—such self/other, truth/deception, control/care, etc.—as not absolute opposites. Instead, they have been treated as complementary and conflicting. In this, understanding one notion depends on and informs knowing its pair. As suggested earlier in this chapter, attempting to know another provides a means of self-knowledge and turning toward oneself a means of knowing another.

As argued, the demands on magicians about how to act are not puzzles to be resolved once and for all. Instead, they are sites of chemistry between different kinds of appreciations. This chemistry pertains to the complex entanglements between authority and empowerment, individuality and joint action, as well as connection and separation.

One benefit of approaching conjuring in this manner is that it provides a basis for acknowledging alternative ways of making sense of a host of practical matters. For instance, conjurors debate questions such as:

- To be considered a 'proper' magician does one need to develop dexterous manual skills or is it possible to rely on manufactured gimmicks?
- Is it wise to foreshadow an intended feat?
- Do magicians need to be proficient in a range of effects or only hone a few?
- Should conjurors portray magic as taking place by them, through them or even to them? Relatedly, should they strive to make the magic appear effortless or strenuous?

- Can a performer gauge the effectiveness of their tricks by taking the visible reactions of others at face value?
- Does understanding the methods for a trick decrease or enhance the sense of wonder associated with witnessing it?
- Is magic a form of artistic self-expression in which the artist's aesthetic judgements should shine through, or is it a form of entertainment in which the audience's judgements are the ones that ultimately count?[14]
- Should beginners imitate their idols or should they seek out their own style?
- When things 'go wrong', is this an opportunity for making an emotional connection with the audience, or a source of disappointment that should be passed over as quickly as possible?

As noted previously, different magicians give different answers to such questions. More than this though, individual magicians can offer opposing counsel at alternative points in time too. In characterizing entertainment magic as *deft contrariwise performance*, the prevalence of clashing responses is not unexpected. Nor does the existence of such advice in itself stand as evidence that some magicians simply do not grasp what they are doing. Instead, the possibility of conflicting counsel can stem from how conjuring entails bringing together the old and the new, the familiar and the unfamiliar, the conventional and the unconventional, and so on.

When conceiving of magic as *deft contrariwise performance*, skill is, in part, the ability to hold together varied ways of assessing what is appropriate. This is another kind of trick that magicians perform. Acting appropriately can be a subtle and fluid undertaking since determinations of what should be done are highly dependent on the sought purposes for performing. Furthermore, any particular purpose—for instance, to entertain; to produce wonder; to inject meaning into life; to reenchant the world; to disaffirm our collective illusions; etc.[15]—can itself be

---

14 Contrast, for instance, Mancha, Hector and Jeremy, Luke. 2006. *3510*. Rancho Cordova, CA: Penguin Magic: 13.
15 For a discussion on the purposes of magic see Burger, Eugene and Neale, Robert E. 1995. *Magic and Meaning*. Seattle, WA: Hermetic P.

questioned for how it involves an interplay of contrary considerations. Determinations of what counts as appropriate action also depend on the varied anticipations, perspectives and identities of those involved, the particulars of performance situations, cultural predispositions, predominant social habits, as well as many other considerations.

As such, the availability of contrasting advice about how magic ought to be performed serves as a basis for debating and assessing. This is particularly important for this art form because of the relative absence of formal institutions for training and accreditation that can serve to establish community-wide standards. On the darker side, the prevalence of contrasting ways of thinking also has the potential to lead to highly evaluative judgements of alternative styles, as well as defensive responses to criticism.[16] In my experience, both of these potentials get realized when conjurors come together during conventions, clubs and online forums. And yet, I have been struck by how magicians respectfully watch each other, share their techniques and even seek out criticism. Learning from one another and teaching one another are central features of collective gatherings. At one level, such behavior is hardly surprising, because being attentive to how other magicians conduct themselves—how they marshal distraction, plan spontaneity, pretend to be natural and so on—helps other magicians to notice what they might not have appreciated about themselves. It also enables individuals to both situate and differentiate themselves in relation to prevailing styles.

Learning From Magic

With this understanding of skill as entailing the interplay of opposing tendencies, I now turn to contrasting the approach to competency development offered in *Performing Deception* with those approaches offered for other domains of activity.

To do so I want to begin with the relation between sensing and knowing. As described in the previous chapters, magic plays up our inclinations to perceive patterns, to adopt the belief that the world exists independently of us, and many other taken-for-granted ways of orientating to our surroundings. Consequently, through learning magic,

---

16 For one practitioner's effort to acknowledge and address defensive reasoning in magic, see Weber, Ken. 2003. *Maximum Entertainment*. Ken Weber Productions.

commonplace ways of understanding the relationship between the senses and knowledge become problematic.

Take sight, for example. The contention that seeing and knowing support each other has widely figured as a theme in the cultural and social analysis of skills acquisition. Roepstorff presented learning to navigate through glaciers and to read brain scans as hard-won enskillments. For such activities, refined vision underpins adept situated action.[17] For O'Connor,[18] sight functioned as a taken-for-granted means of receiving sensory inputs that enabled glassblowers to gain nuanced types of focal and subsidiary awareness.

Learning, in my case, certainly entailed the refinement of visual-motor skills (for instance, finger positioning) through assessing actions (spreading, cutting, bending, placing and lifting cards) against intended outcomes. However, what has also come to the fore has been the complex and sometimes indeterminate relationship between seeing and knowing. In the practices surveyed in previous chapters, seeing could not straightforwardly be taken as knowing (for instance: knowing *whether* physical manipulations are detectable; knowing *that* someone is being truthful; knowing *how* reliably a visual effect can be repeated). Knowing, too, fostered questioning of what takes place in seeing. This happened, for instance, in relation to what was not made visible in instructional videos and to the alluring seductions of gazing into a mirror when you know what to look for.

In other words—as part of my development—I came to know, to realize I did not know, to wonder what I could know, and to doubt what I thought I knew. In doing so, I experienced a growing uneasiness about the intelligibility and reliability of the visual, even though in many other respects I treated visual perception as unproblematic.[19]

In such ways, as I engaged in conjuring for others, the world transformed into a kind of conjuring.

As a result, definitions that depict learning as a process of matching 'this to that'—for instance, error detection against expected outcomes,[20]

---

17 Roepstorff, A. 2007. 'Navigating the Brainscape'. In: *Skilled Visions: Between Apprenticeship and Standards*, C. Grasseni (Ed.). Oxford: Berghahn Books: 191–206.
18 O'Connor, E. 2005. 'Embodied Knowledge', *Ethnography*, 6: 183–204.
19 A troubling for which many historical parallels could be made; see Clark, Stuart. 2007. *Vanities of the Eye*. Oxford: Oxford University Press.
20 Argyris, C. 1995. 'Action Science and Organizational Learning', *Journal of Managerial Psychology*, 10(6): 20–26.

or of linking stimulus to responses,[21] or of disciplining errors to achieve greater skillfulness[22]—only capture some of the dynamics surveyed in previous chapters. My own fraught learning involved a maturing hesitancy about my claims to individual agency and control, even as I became defter in physically moving cards and socially interacting with audiences. Learning was a process undertaken concerning imaginary or actual others, yet others with a shifting status. Others were (un)available to me in relation to our shared experiences, our different experiences and, importantly, my growing hesitancy regarding whether we had similar or different experiences.

As I have come to understand conjuring, learning it entails adeptly acting in between certainty and uncertainty, as well as the possibilities for affirmation and not. In this way, learning involved what anthropologist Tim Ingold coined as an 'education of attention'.[23] That is to say, it involved sensitization of the perceptual system. However, educating attention entailed an *unsettling* of perception too, not simply honing it. This unsettling took place at two levels: one, making sense of specific sensory experiences (what was seen in looking in this mirror, watching that video, etc.) and, two, making sense of the sensory capacities in general (the possibilities for discernment given the fallibilities of human perception).

Taking these points together with themes from previous chapters, learning magic has entailed developing a receptiveness to *movement*; that is, an ability to to-and-fro between:

- particular situated events and general descriptions;
- the reliance on others' accounts and the questioning of them;
- the credence given to and the distancing from sensory experiences;

---

21 Lachman, S.J. 1997. 'Learning is a Process', *The Journal of Psychology*, 131(5): 477–480. https://doi.org/10.1080/00223989709603535.

22 Downey, G., Dalidowicz, M., and Mason, P.H. 2015. 'Apprenticeship as Method', *Qualitative Research*, 15(2): 183–200. https://doi.org/10.1177/1468794114543400.

23 In doing so, Ingold adopted James Gibson's term, see Ingold, T. 2001. 'From the Transmission of Representations to the Education of Attention'. In: *The Debated Mind: Evolutionary Psychology Versus Ethnography*, H. Whitehouse (Ed.). London: Bloomsbury Academic: 113–154. https://doi.org/10.4324/9781003086963-7.

- resting with what one has learnt and seeking to unlearn;
- treating other people's experiences as distinct as well as similar to one's own;
- losing oneself in play and being aware that one is playing.

Part of the demand of performing magic is being able to adapt to and shift between such orientations. This can entail recognizing what is readily accessible; appreciating what requires refined judgement; perceiving with foreknowledge; disregarding foreknowledge; watching what is demonstrated, and imagining what is not shown. Undertaking such acts can also entail moving between different working theories regarding how we know ourselves and each other. I refer to the development of the ability to move between certainty and uncertainty, as well as affirmation and its unattainableness, as *trick learning*.

The comments in the previous paragraphs are not just relevant to the practical task of learning magic. They apply to the account given in *Performing Deception*. The analysis in these pages—which is to say, the relationship between the teller and the told—is caught within the kinds of tensions set out. Notably, I have used my own sense-making as the principal way into considering the basis for sense-making. This tension-ridden situation is hardly unique to *Performing Deception*, as any inquiry of reasoning faces a basic conundrum of how to examine the means it uses to undertake that examination.[24] There is then a second-order challenge regarding how to communicate the questioning of commonplace reasoning to readers such as yourself. As I have argued, the activity of magic makes relevant a third dimension of challenge: the fallibility of commonplace reasoning and perception. Rather than somehow escaping these challenging conditions, I have sought to convey my emerging understanding as a novice as a way into appreciating how notions of commonsense and sense-making are at stake in the doings of magic.

There is another important manner in which the analysis in these pages is caught within the kinds of tensions set out. While learning magic has been conceived as a process of relating lived concrete

---

24 A theme taken up in Ten Have, Paul. 2004. *Understanding Qualitative Research and Ethnomethodology*. London: Sage. https://doi.org/10.4135/9780857020192.

experiences, abstract concepts and theories, active experimentation, as well as observations and reflections, what *Performing Deception* has provided is a set of abstractions and reflections. As a reduction of worldly encounters and practical abilities into a written account, this book has not been able to somehow convey embodied experiences and actions fully. What it has been able to do is provide an intellectual guide for appreciating the illusionary nature of our everyday ways of making sense of the world. Part of the trick of crafting this book has been to offer plausible descriptions and arguments that build shared understandings of learning, despite the limitations in what is presented.

To acknowledge how telling and obscuring come together in this manner is to further open up to what learning entails. This is not a steady progression from ignorance to knowledge or from ineptitude to proficiency, but an ongoing process of coming into and out of tension and paradox.

Index

accountability 80
Agnes 78
Allen, Jon 149
Aronson, Simon 48
attention and challenge
  interpersonal relation in a virtual setting 161
autobiographies 18
  as a form of writing 126, 145
  as performative writing 126
  truth and deception in contemporary accounts 130

Baxter, Leslie 14
beginner's mind 4
Being and Nothingness 36
Billig, Michael 105–106
Blaine, David 7, 79, 143, 209
Bogen, David 135
Book of Secrets, The 142
Braué, Frederick 76
Brown, Derren 7, 69–72, 75, 77, 100, 137–142, 144, 150, 156, 196, 204
Buber, Martin 168

Carbonaro Effect 81–82
care
  and control 160
  caring for objects 157
  deception as 154
  projection 169
  reception 169
Chinn, Mielin 80
Cohen, Michael D 92, 136–137, 197, 207
Confessions of a Conjuror 140
Confidences d'un Prestidigitateur 127
Connection 149
connection-disconnection 16–17, 46, 72, 179
contrariwise
  definition 13
control 52
  challengers 49, 151
  cooperation 52
  making an emotional connection to others 148
  molding spectators' imagination 47
cooperation 52, 60
Copperfield, David 7, 176
Corrieri, Augusto 19, 86
Cruel Tricks for Dear Friends 130

Damkjaer, Camilla 22
DaOrtiz, Dani 13, 90, 115
deception 11–12, 16, 69, 104, 132, 159, 178
  and control 47
  and discernment 52, 71, 80
  and self-presentation 71
  and trust 133–134, 139
  care 154, 168
  displaying 130, 132, 138, 142
  naturalness 79
  of magicians 70
  shame 145
  sincere liar 135
deft
  definition 9
deft contrariwise performance
  characterizing entertainment magic as 8, 183
de Grisy, Edmond 128
De Man, Paul 144
DeNora, Tia 108
Dewey, John 14
dialectics 15, 19, 183, 185
directive trajectory 54
Dolezal, Luna 37
dynamic interplay

magic as 15, 50, 53, 67, 183
Dynamo 142, 145

Earl, Benjamin 86–87, 93, 95, 97, 100
education of attention 187
effects
  defintion 30
Erdanese, S.W. 75
ethics of care 153
ethnomethodology 6, 16, 68, 108
Experiencing the Impossible 107
experiential learning
  learning magic as 4, 182, 189
Expert at the Card Table, The 75

Fitzkee, Dariel 47
Frisch, Ian 125
Fulves, Karl 24

Gallagher, Shaun 39, 68, 72, 199
Garfinkel, Harold 6, 25, 78
Glaser, Donald 85
Goffman, Erving 9
Goodwin, Charles 108
Goto-Jones, Chris 177
Grice, Paul 52

Haraway, Donna 4, 73
Hartling, Pit 49
Helge, Thun 49
Houdini, Harry 129
How to Play with your Food 132
Hugard, Jean 41, 75–76, 86, 155, 201

ignorance 3, 15
  as strength 4
Ingold, Tim 187
instructions
  audio-visual learning 41
  correspondence 32
  envisaging 36
  managing their relevance 24
interpersonal communication
  dialectical approach to 14
intersubjectivity 16, 72

Jay, Joshua 44

Jerx, The 81
Jillette, Penn 7, 13, 135, 137, 143, 146
Jillette, Penn & Teller 133, 206
Jones, Graham 30, 128, 176

Kolb, David 4
Kuhn, Gustav 107

Leddington, Jason 48, 175
Liberman, Kenneth 62
Livingston, Eric 25, 32, 181
Lynch, Michael 108, 135

magic
  as deft contrariwise performance 13, 15, 74, 183
  care 152, 158
  definition 8, 175
  doubleness 12, 43, 73
  ethical limits 12, 134, 149
  imitation 95, 99
  modern 17, 77
  opposing advice 106, 156, 183
  pathways for training 23
  rationality 177
  reciprocal action 2, 66
  revealing secrets x, 146, 155
  topics and resources 19, 173
  via Zoom 162
Magic Circle ix, xi, 100, 203
Magic is Dead 125
Magic's Reason 176
Mangan, Michael 126, 129
Marchand, Trevor 33
Martin, Aryn 153
material objects 7
  accountability 79
  as passive 84
  care 158
  mirrors 86
  naturalness 77
meta-magic 163
method
  defintion 30
Mol, Annemarie 7
Montgomery, Barbara 14

Mulholland, John 11

Nardi, Peter 47
natural attitude 73–74
naturalness 17, 96
  deceiving through 77
  emotional engagement 91
  feeling right vs. looking right 87, 93, 95, 97, 100
  mirror practice 82
  relating to the material world 84
  what counts as natural 75
Noddings, Nel 169
norms 66
North, Oliver
  testimony of 135
Nothing is Impossible 142

O'Connor, Erin 186
ordinariness 79
Ortiz, Darwin 6, 11, 49, 56, 60, 76, 96, 101–104, 106, 156, 205

Palmore, Steve 150–151, 155, 159, 164
perception
  fallible 2, 18, 100, 186–187
  magicians 103
performance
  definition 10–11, 14, 16–17, 19, 23–24, 206
Performing Dark Arts 126
philosophy of mind 39, 72
  simulation 38
Pickering, Andy 7, 84–85, 206
Pollner, Melvin 110
prestigious imitation 90
proficiency
  accounting for perception 108
  learned incompetency 102
Puig, Maria de la Bellacasa 157

Real Deck Switches 87, 93, 95, 97, 100
Robert-Houdin, Jean-Eugène 127, 144
Robinson, William 80
Roepstorff, Andreas 186
Rolfe, Charles 50, 81, 197, 207
Roth, Ben 144

Royal Road to Card Magic, The 41–42, 75

Sankey, Jay 81
Sartre, Jean-Paul 36–37, 198, 207
Schutz, Alfred 73
Science of Magic 108
seeing and knowing
  complex relationship between 185
  moving between certainty and uncertainty 188
self-other study
  conceiving learning magic as 2, 22
  knowing the audience 44
self-presentation 71, 90
Self-Working Card Tricks 24–25, 46
  Lazy Magician, The 34
  No-Clue Discovery 26, 37, 45
skill 17, 22, 24–25, 33, 45, 56, 70, 77, 91, 100, 111, 142, 151, 155, 182, 184
sleight of hand 41, 51, 90, 99–100, 111
Smith, Wally 6, 11, 24, 26, 48, 77, 79, 109, 127, 158, 208
Soo, Chung Ling 80
Steinmeyer, Jim 129
Strong Magic 49, 101
Suzuki, Shunryu 4
sympoiesis
  conjuring as an activity of 73

Tamariz, Juan 13
Teller 7, 13, 135, 137, 143, 146
The Magic of Social Life 163, 206
The Unmasking of Robert-Houdin 129
Trade of Tricks 30
transcribed group performances 53, 109
trick learning 188
Tricks of the Mind 137
Trump, Donald J 135–136

Vanishing Inc 30, 44, 49, 80, 87, 102, 149–150, 197–198, 200–201, 204

Wacquant, Loïc 12
Weber, Michael 147
Weber, Mike 155
Wetherell, Margaret 96

Zahavi, Dan 72

Bibliography

Adler, J. 1997. 'Lying, Deceiving, or Falsely Implicating'. *The Journal of Philosophy*, 94(9): 435–452. https://doi.org/10.2307/2564617.

Allen, Jon. 2019, June 19. *Day of Magic Presentation*. Leamington Spa.

Allen, Jon. 2013. *Connection*. Las Vegas, NV: Penguin Magic.

Allen, Jonathan and O'Reilly, Sally. 2009. *Magic Show*. London: Hayward Publishing.

Allen, Jonathan. 2007. 'Deceptionists at War', *Cabinet* (Summer), 26. http://www.cabinetmagazine.org/issues/26/allen.php.

Ameel, Lieven and Tani, Sirpa. 2012. 'Everyday Aesthetics in Action: Parkour Eyes and the Beauty of Concrete Walls', *Emotion, Space and Society*, 5: 164–173.

American Heritage® Dictionary of the English Language (Fifth Edition). 2011. https://www.thefreedictionary.com/contrariwise

Argyris, C. 2006. *Reasons and Rationalizations*. Oxford: Oxford University Press. https://doi.org/10.1093/acprof:oso/9780199268078.001.0001.

Argyris, C. 1999. *On Organizational Learning*. Cambridge: Blackwell.

Argyris, C. 1995. 'Action Science and Organizational Learning', *Journal of Managerial Psychology*, 10(6): 20–26.

Armstrong, Jon. 2019. *Insider* (16 December). https://www.vanishingincmagic.com/insider-magic-podcast/

Aronson, Simon. 1990. *The Illusion of Impossibility*. [n.p.]: Simon Aronson.

Arundale, Robert B. 2010. 'Constituting Face in Conversation: Face, Facework, and Interactional Achievement', *Journal of Pragmatics*, 42: 2078–2105. https://doi.org/10.1016/j.pragma.2009.12.021.

Atkinson, P. 2013. 'Blowing Hot', *Qualitative Inquiry*, 19(5): 397–404. https://doi.org/10.1177/1077800413479567.

Atkinson, Paul and Silverman, David. 1997. 'Kundera's Immortality', *Qualitative Inquiry*, 3: 304–325.

Atkinson, Paul, Watermeyer, Richard and Delamont, Sara. 2013. 'Expertise, Authority and Embodied Pedagogy: Operatic Masterclasses', *British Journal* of *Sociology of Education*, 34(4): 487–503. https://doi.org/10.1080/01425692.2012.723868.

Avner Insider, 2019: https://www.vanishingincmagic.com/insider-magic-podcast/.

Bakhtin, Mikhail. 1981. *The Dialogic Imagination*. Austin, TX: University of Texas Press.

Baxter, Leslie A. and Montgomery, Barbara M. 1996. *Relating: Dialogues and Dialectics*. London: Guilford.

Beckman, Karen. 2003. *Vanishing Women: Magic, Film, Feminism*. Durham, NC: Duke University Press. https://doi.org/10.1215/9780822384373.

Bell, Karl. 2012. *The Magical Imagination*. Cambridge: Cambridge University Press.

Billig, M. 1996. *Arguing and Thinking*. Cambridge: Cambridge University Press.

Billig, M., S. Condo, D. Edwards, M. Gane, D. Middleton and A. Radley. 1989. *Ideological Dilemmas*. London: Sage.

Blackstone, Harry. 1977. *Blackstone's Secrets of Magic*. North Hollywood, CA: Wilshire.

Blaine, David. 2002. *Mysterious Stranger*. New York: Villard.

Bordo, Susan. 1993. *Unbearable Weight*. London: University of California Press.

Brown, Derren and Swiss, Jamy Ian. *A Conversation in Two Parts: Part I*. 2003, June 29. http://honestliar.com/fm/works/derren-brown.html.

Brown, Derren. 2021. *Bristol Society of Magic—Centenary Celebration: An Evening with Derren Brown* (Bristol), 3 May.

Brown, Derren. 2010. *Confessions of a Conjuror*. London: Channel 4 Books.

Brown, Derren. 2006. *Tricks of the Mind*. London: Channel 4 Books.

Brown, Derren. 2003. *Absolute Magic: A Model for Powerful Close-Up Performance* (Second edition). London: H&R Magic Books.

Bruns, L. C. and Zompetti, J. P. 2014. 'The Rhetorical Goddess: A Feminist Perspective on Women in Magic', *Journal of Performance Magic*, 2(1). https://doi.org/10.5920/jpm.2014.218.

Buber, Martin. 2018. *I and Thou*. London: Bloomsbury Academic.

Burger, Eugene and Neale, Robert E. 1995. *Magic and Meaning*. Seattle, WA: Hermetic Press.

Burger, Eugene. [n.d.], *Audience Involvement… A Lecture*. Asheville, NC: Excelsior!! Productions.

Butler, J. 2007. *Gender Trouble: Feminism and the Subversion of Identity*. London: Routledge Classics.

Charles, Rolfe. 2020. 'Theatrical Magic and the Agenda to Enchant the World', *Social & Cultural Geography*, 17(4): 574–596. https://doi.org/10.1080/14649365.2015.1112025.

Chinn, Mielin 2019. 'Race Magic and the Yellow Peril', *The Journal of Aesthetics and Art Criticism*, 77(4): 423–433.

Clark, Stuart. 2007. *Vanities of the Eye*. Oxford: Oxford University Press. https://doi.org/10.2752/175183409x12550007730345.

Clarke, S. 1999. 'Justifying Deception in Social Science Research', *Journal of Applied Philosophy*, 16(2): 151–166. https://doi.org/10.1111/1468-5930.00117;

Clark, Stuart. 1997. *Thinking with Demons*. Oxford: Oxford University Press.

Clifford, Peter. 2020, January 12. *A Story for Performance*. Lecture notes from presentation at The Session. London.

Close, Michael. [2003] 2013. 'The Big Lie'. In: *Magic in Mind: Essential Essays for Magicians*, Joshua Jay (Ed.). Sacramento: Vanishing Inc.

Cohen, Erik and Cohen, Scott A. 2012. 'Authentication: Hot and Cold' *Annals of Tourism Research*, 39(3): 1294–1314. https://doi.org/10.1016/j.annals.2012.03.004.

Cohen, Michael D. 2019. Committee on Oversight and Reform U.S. House of Representatives, February 27. https://www.theguardian.com/us-news/2019/feb/28/trump-says-cohen-lied-testimony-congress

Colie, Rosalie. 1966. *Paradoxia Epidemica: The Renaissance Tradition of Paradox*. Princeton, NJ: Princeton University Press.

Collins, H. and Evans, R. 2002. *Rethinking Expertise*. Chicago, IL: University of Chicago Press. https://doi.org/10.7208/chicago/9780226113623.001.0001.

Coppa, Francesca, Hass, Lawrence and Peck, James (Eds) 2008. *Performing Magic on the Western Stage*. London: Palgrave MacMillan: 8. https://doi.org/10.1057/9780230617124.

Corrieri, Augusto. 2018. '"What Is This..."': Introducing Magic and Theatre, *Platform*, 12(2): 12–17.

Corrieri, Augusto. 2016. 'An Autobiography of Hands', *Theatre, Dance and Performance Training*, 7(2): 283–229. https://doi.org/10.1080/19443927.2016.1175501

Corrigan, B. J. 2018. '"This Rough Magic I here Abjure": Performativity, Practice and Purpose of the Bizarre', *Journal of Performance Magic*, 5(1). https://doi.org/10.5920/jpm.2018.05.

Coulter, J. and Parsons, E.D. 1991. 'The Praxiology of Perception', *Inquiry*, 33: 251–272.

Damkjaer, Camilla. 2016. *Homemade Academic Circus*. Winchester: iff.

Danek, Amory H., Fraps, Thomas, von Müller, Albrecht, Grothe, Benedikt and Öllinger, Michael. 2013. 'Working Wonders? Investigating Insight with Magic Tricks', *Cognition, 130*(2): 174–185. http://dx.doi.org/10.1016/j.cognition.2013.11.003.

DaOrtiz, Dani. 2018. *Working at Home*. GrupoKaps.

DaOrtiz, Dani. 2017. *Penguin Dani DaOrtiz LIVE ACT*. https://www.penguinmagic.com/p/11142

Dawson, Sheila. 1961. '"Distancing" as an Aesthetic Principle', *Australasian Journal of Philosophy. 39*: 155–174.

De Man, Paul. 1979. *Allegories of Reading*. New Haven: Yale University Press.

Dean, E. 2018. 'The End of Mindreading', *Journal of Performance Magic, 5*(1). https://doi.org/10.5920/jpm.2018.04

DeNora, Tia. 2014. *Making Sense of Reality*. London: Sage. https://dx.doi.org/10.4135/9781446288320.

DePaulo, B.M., Kashy, D. A., Kirkendol, S. E., Wyer, M. M., and Epstein, J. A. 1996. 'Lying in Everyday Life', *Journal of Personality and Social Psychology, 70*: 979–995. https://doi.org/10.1037/0022-3514.70.5.979.

Devant, D. and Maskelyne, N. 1912. *Our Magic*. London: George Routledge & Sons.

Dewey, John. 1934. *Art as Experience*. New York: Perigee Books.

Dolezal, Luna. 2012. 'Reconsidering the Look in Sartre's: *Being and Nothingness*', *Sartre Studies International, 18*(1): 18. https://doi.org/10.3167/ssi.2012.180102.

Downey, G., Dalidowicz, M., and Mason, P.H. 2015. 'Apprenticeship as Method', *Qualitative Research, 15*(2): 183–200. https://doi.org/10.1177/1468794114543400.

During, Simon. 2002. *Modern Enchantments*. London: Harvard University Press. https://doi.org/10.4159/9780674034396.

Dynamo. 2017. *Dynamo: The Book of Secrets*. London: Blink Publishing.

Dynamo. 2012. *Nothing Is Impossible*. London: Ebury Press.

Earl, Ben. 2020, July 18. *Deep Magic Seminar*.

Earl, Benjamin. 2018. *Roleplayer*. Sacramento, CA: Benjamin Earl & Vanishing Inc.

Earl, Benjamin. [n.d.]. *Real Deck Switches*. Sacramento, CA: Vanishing Inc.

Ekroll, Vebjørn Bilge Sayim, and Wagemans, Johan. 2017. 'The Other Side of Magic', *Perspectives on Psychological Science, 12*(1): 91–106. tps://doi.org/10.1177/1745691616654676.

Elderfield, Tom. 2019, June 17. *The Insider*. https://www.vanishingincmagic.com/blog/the-insider-tom-elderfield

Erdnase, S.W. 1955. *The Expert at the Card Table*. Mineola, NY: Dover.

Evans, J., Davis, B. and Rich, E. 2009. 'The Body Made Flesh: Embodied Learning and the Corporeal Device', *British Journal of Sociology of Education*, 30(4): 391–406. https://doi.org/10.1080/01425690902954588.

Fitzkee, Dariel. 1945. *Magic by Misdirection*. Provo, UT: Magic Book Productions.

Fligstein, Neil. 2001. 'Social Skill and the Theory of Fields', *Sociological Theory*, 19(2): 105–125. https://doi.org/10.1111/0735-2751.00132.

Francis, David and Hester, Stephen. 2004. *An Invitation to Ethnomethodology: Language, Society and Interaction*. London: Sage: 7.

Frisch, Ian. 2019. *Magic Is Dead: My Journey into the World's Most Secretive Society of Magicians*. New York: Dey St.

Fulves, K. 1976. *Self-Working Card Tricks*. New York: Dover.

Gallagher, S. 2005. 'How the Body Shapes the Mind'. In: *Between Ourselves: Second-Person Issues in the Study of Consciousness*, Evan Thompson (Ed.). Oxford: Oxford University Press.

Gallagher, S. 2001. 'The Practice of Mind', *Journal of Consciousness Studies*, 8(5–7): 83–108.

Gallagher, Shaun and Zahavi, Dan. 2007. *The Phenomenological Mind: An Introduction to Philosophy of Mind and Cognitive Science*. London: Routledge: 187.

Garcia, Frank. 1972. *Million Dollar Cardsecrets*. New York: Million Dollar Productions.

Garfinkel, Harold. 2002. *Ethnomethodology's Program*. Oxford: Rowman and Littlefield.

Garfinkel, Harold. 1984. *Studies in Ethnomethodology*. Cambridge: Polity.

Gilligan, Carol. 1982. *In a Different Voice*. Cambridge, MA: Harvard University Press.

Girton, George D. 1986 'Kung Fu'. In: *Ethnomethodological Studies of Work*, H. Garfinkel (Ed.). London: Routledge & Kegan Paul.

Goffman, E. 1959. *The Presentation of Self in Everyday Life*. Harmondsworth: Penguin.

Goffman, Erving. 1956. *The Presentation of Self in Everyday Life*. New York: Doubleday: 8.

Goldman, A.I. 2002. 'Simulation Theory and Mental Concepts'. In: *Simulation and Knowledge of Action*, Dokic, J. and Proust, J. (Eds). Amsterdam: John Benjamins: 35–71. https://doi.org/10.1075/aicr.45.02gol.

Goodwin, Charles. 2017. *Co-Operative Action*. Cambridge: Cambridge University Press.

Goodwin, C. 2003. 'Pointing as Situated Practice'. In: *Pointing: Where Language, Culture and Cognition Meet*, S. Kita (Ed.). Mahwah, N.J.: Lawrence Erlbaum.

Goodwin, C. 1995. 'Seeing in Depth', *Social Studies of Science*, 25: 237–274.

Goodwin, Charles. 1994. 'Professional Vision', *American Anthropologist*, 96(3): 606– 633.

Goodwin, Marjorie Harness and Cekaite, Asta. 2012. 'Calibration in Directive/ Response Sequences in Family Interaction', *Journal of Pragmatics*, 46(1): 122–138. http://dx.doi.org/10.1016/j.pragma.2012.07.008.

Goto-Jones, Chris. 2016. *Conjuring Asia: Magic, Orientalism, and the Making of the Modern World*. Cambridge University Press. https://doi.org/10.1017/cbo9781139924573.

Granrose, John. 2021. *The Archetype of the Magician*. Agger: Eye Corner Press

Greenbaum, Harrison. *The Insider*. 18 November 2019. https://www.vanishingincmagic.com/blog/the-insider-harrison-greenbaum.

Grice, Paul. 1989. *Studies in the Way of Words*. Cambridge, MA: Harvard University Press.

Haraway, Donna. 2016. *Staying with the Trouble*. Durham, NC: Duke University Press.

Haraway, Donna. 1988. 'Situated Knowledges: The Science Question in Feminism and the Privilege of Partial Perspective', *Feminist Studies* (Autumn), 14(3): 575–599.

Harris, Paul and Mead, Eric. *The Art of Astonishing*. [n.p.]. Multimedia A-1.

Hartling, Pit. [2003] 2013. 'Inducing Challenges'. In: *Magic in Mind: Essential Essays for Magicians*, Joshua Jay (Ed.). Sacramento: Vanishing Inc.: 105–112.

Hass, Lawrence. (Ed.) 2010. *Gift Magic: Performances That Leave People with a Souvenir*. Theory and Art of Magic Press.

Heath, Christian. 1984. 'Talk and Recipiency: Sequential Organization in Speech and Body Movement'. In: *Structures of Social Action: Studies in Conversational Analysis*, J.M. Atkinson and J. Heritage (Eds). Cambridge: Cambridge University Press: 247–265. https://doi.org/10.1017/cbo9780511665868.017.

Held, Virginia. 1993. *Feminist Morality: Transforming Culture, Society, and Politics*. Chicago, IL: University of Chicago Press

Hendriks, Ruud. 2012. 'Tackling Indifference—Clowning, Dementia, and the Articulation of a Sensitive Body', *Medical Anthropology*, 31(6): 459–476. https://doi.org/10.1080/01459740.2012.674991

Heritage, John. 1984. *Garfinkel and Ethnomethodology*. Cambridge: Polity Press.

Hill, Annette. 2010. Paranormal Media: Audiences, Spirits, and Magic in Popular Culture. London: Routledge: 142–149. https://doi.org/10.4324/9780203836392.

Hochschild, A.R., 2003. *The Managed Heart: Commercialization of Human Feeling*. London: University of California Press. https://doi.org/10.1525/9780520930414.

Hopkins, Charles. 1978. *Outs, Precautions and Challenges for Ambitious Card Workers*. Calgary: Micky Hades.

Houstoun, W. and Thompson, S. 2021. *Video Chat Magic*. Sacramento, CA: Vanishing.

Houstoun, Will and Thompson, Steve. 2020, June 7. 'Are You a Prioritisationalist?', *Video Chat Magic*. See https://videochatmagic.substack.com/p/are-you-a-prioritisationalist

Hugard, Jean and Braué, Frederick. 2015. The Royal Road to Card Magic (Video Edition). London: Foulsham.

Hunt, Jennifer and Manning, Peter K. 1991. 'The Social Context of Police Lying', *Symbolic Interaction*, 14(1): 51–70. https://doi.org/10.1525/si.1991.14.1.51.

Ingold, T. 2001. 'From the Transmission of Representations to the Education of Attention'. In: *The Debated Mind: Evolutionary Psychology Versus Ethnography*, H. Whitehouse (Ed.). London: Bloomsbury Academic: 113–154. https://doi.org/10.4324/9781003086963-7.

Ivinson, G. 2012. 'The Body and Pedagogy: Beyond Absent, Moving Bodies in Pedagogic Practice', *British Journal of Sociology of Education*, 33(4): 489–506. https://doi.org/10.1080/01425692.2012.662822.

Jay, Joshua. 2020. January 9. *Presentation at The Session*. London.

J. Kitzinger and Barbour, R. 1999. 'Introduction'. In: *Developing Focus Group Research*, R. Barbour and J. Kitzinger (Eds). London: Sage.

Jay, J. 2016. 'What do Audiences Really Think?' MAGIC (September): 46–55. https://www.magicconvention.com/wp-content/uploads/2017/08/Survey.pdf

Jay, Joshua (Ed). 2013. *Magic in Mind: Essential Essays for Magicians*. Sacramento: Vanishing Inc.

Johns, Christopher. 2009. *Becoming a Reflective Practitioner* (Third Edition). Oxford: Wiley-Blackwell.

Jon Armstrong, 2019. *Insider*. https://www.vanishingincmagic.com/insider-magic-podcast/

Jones, Graham and Shweder, Lauren. 2003. 'The Performance of Illusion and Illusionary Performatives: Learning the Language of Theatrical Magic', *Journal of Linguistic Anthropology*, 13(1): 51–70. https://doi.org/10.1525/jlin.2003.13.1.51.

Jones, Graham M. 2018. *Magic's Reason*. Chicago: Chicago University Press. https://doi.org/10.7208/chicago/9780226518718.001.0001.

Jones, Graham. 2012. 'Magic with a Message', *Cultural Anthropology,* 27(2): 193–214. https://doi.org/10.1111/j.1548-1360.2012.01140.x.

Jones, Graham. 2011. *Trade of the Tricks*. London: University of California Press. https://doi.org/10.1525/california/9780520270466.001.0001.

Kaps, Fred. 1973. *Lecture Notes*. London: Ken Brookes' Magic Place.

Kendon, A. 1986. 'Some Reasons for Studying Gesture', *Semiotica,* 62: 3–28.

Kestenbaum, David. 2017, June 30. 'The Magic Show—Act Two', *The American Life*. https://www.thisamericanlife.org/619/the-magic-show/act-two-31

Kirsh, David. 2006. 'Distributed Cognition', *Pragmatics & Cognition,* 14(2): 249–262. https://doi.org/10.1075/pc.14.2.06kir.

Kittay, Eva Feder. 1999. *Love's Labor*. London: Routledge.

Kolb, D. A. 2015. *Experiential Learning* (Second Edition). Saddle River, NJ: Pearson Education.

Kuhn, G. 2019. *Experiencing the Impossible*. Cambridge, MA: MIT Press. https://doi.org/10.7551/mitpress/11227.001.0001.

Kuhn, G., Caffaratti, H., Teszka, R. and Rensink, R.A. 2016. 'A Psychologically-Based Taxonomy of Misdirection'. In: *The Psychology of Magic and the Magic of Psychology* (November), Raz, A., Olson, J. A. and Kuhn, G. (Eds). https://doi.org/10.3389/fpsyg.2014.01392.

Kuhn, Gustav, Tatler, Benjamin W. and Cole, Geoff G. 2009. 'Look Where I Look', *Visual Cognition,* 17(6/7): 925–944. https://doi.org/10.1080/13506280902826775.

Kuhn, Gustav, Teszka, Robert, Tenaw, Natalia and Kingstone, Alan. 2016. 'Don't Be Fooled! Attentional Responses to Social Cues in a Face-to-Face and Video Magic Trick Reveals Greater Top-Down Control for Overt than Covert Attention', *Cognition,* 146: 136–142. https://doi.org/10.1016/j.cognition.2015.08.005.

Kuntsman, Adi and Stein, Rebecca. 2015. *Digital Militarism*. Stanford: Stanford University Press. https://doi.org/10.4135/9781473936676.

Lacan, J. 1977. *Ecrits: A Selection* (trans. Alan Sheridan). New York: Norton.

Lachman, S.J. 1997. 'Learning is a Process', *The Journal of Psychology,* 131(5): 477–480. https://doi.org/10.1080/00223989709603535.

Lamont, Peter. 2013. *Extraordinary Beliefs*. Cambridge: Cambridge University Press. https://doi.org/10.1017/CBO9781139094320.

Lamont, Peter. 2009. 'Magic and the Willing Suspension of Disbelief'. In *Magic Show,* Jonathan Allen and Sally O'Reilly (Eds). London: Hayward Publishing: 30.

Lamont, Peter. 2006. 'Magician as Conjuror', *Early Popular Visual Culture,* 4(1): 21–33.

Lamont, Peter and Wiseman, Richard. 1999. *Magic in Theory*. Hatfield: University of Hertfordshire Press.

Landman, Todd. 2020. 'Making it Real'. In: *The Magiculum II*, T. Landman (Ed.). [n.p.]: Todd Landman.

Landman, Todd. 2018. 'Academic Magic: Performance and the Communication of Fundamental Ideas', *Journal of Performance Magic*, 5(1). https://doi.org/10.5920/jpm.2018.02.

Laurier, Eric. 2004. 'The Spectacular Showing: Houdini and the Wonder of Ethnomethodology', *Human Studies, 27*: 385–387. https://doi.org/10.1007/s10746-004-3341-5.

LeClerc, Eric. 2019. *Insider 30 September*. https://www.vanishingincmagic.com/insider-magic-podcast/

Lecoq, Jacques. 2000. *The Moving Body*. London: Bloomsbury.

Leddington, Jason. 2020, May 28. 'Savouring the Impossible', *Aesthetics Research Centre Online Seminar*. http://aesthetics-research.org/archive/2020/leddington/.

Leddington, Jason. 2016. 'The Experience of Magic', *The Journal of Aesthetics and Art Criticism*, 74(3): 255. https://doi.org/10.1111/jaac.12290.

Leeder, Murray. 2010. 'M. Robert-Houdin Goes to Algeria', *Early Popular Visual Culture*, 8(2): 209–225. https://doi.org/10.1080/17460651003688113.

Liberman, Kenneth. 2013. *More Studies in Ethnomethodology*. Albany, NY: State University of New York Press: 108.

Liberman, Kenneth. 2007. *Dialectical Practice in Tibetan Philosophical Culture*. London: Rowman & Littlefield.

Lynch, Michael and Bogen, David. 1996. *The Spectacle of History*. London: Duke University Press.

Livingston, Eric. 2008. *Ethnographies of Reason*. London: Routledge. https://doi.org/10.4324/9781315580555

Luhrmann, Tanya M. 1989. 'The Magic of Secrecy', *Ethos*, 17(2): 131–165.

Lynch, Michael. 2013. 'Seeing Fish'. In: *Ethnomethodology at Play*, P. Tolmie and M. Rouncefield (Eds). London: Routledge: 89–104.

Magic Circle. 2017. *Guide to Examinations* (November). London: Magic Circle. https://themagiccircle.co.uk/images/The-Magic-Circle-guide-to-examinations.pdf

Magic, Charlatanry and Skepticism, Comments at SOMA Magic & Creativity Webinar. https://scienceofmagicassoc.org/blog/2021/4/29/magic-charlatanry-skepticism-webinar-cd6cy).

Malvern, Jack. 2019, January 2. 'Magicians Accused of Casting Pseudoscience Spell on Audiences', *The Times*.

Mancha, Hector and Jeremy, Luke. 2006. *3510*. Rancho Cordova, CA: Penguin Magic.

Mangan, Michael. 2017. 'Something Wicked: The Theatre of Derren Brown'. In: *Popular Performance, Adam Ainsworth, Oliver Double and Louise Peacock* (Eds). London: Bloomsbury.

Mangan, Michael. 2007. *Performing Dark Arts: A Cultural History of Conjuring*. Bristol: Intellect.

Marchand, Trevor H.J. 2010. 'Embodied Cognition and Communication', *Journal of the Royal Anthropological Institute*, 16: S112. https://doi.org/10.1111/j.1467-9655.2010.01612.x

Marchand, Trevor H.J. 2008. 'Muscles, Morals and Mind: Craft Apprenticeship and the Formation of Person', *British Journal of Educational Studies*, 56(3): 245–271. https://doi.org/10.1111/j.1467-8527.2008.00407.x.

Martin, Arrn, Myers, Natasha and Viseu, Ana. 2015. 'The Politics of Care in Technoscience', *Social Studies of Science*, 45(5): 635. https://doi.org/10.1177/0306312715602073

Maskelyne, Neil and Devant, David. 1911. *Our Magic*. London: George Routlege & Sons.

Mauss, M. 1973 [1934]. 'Techniques of the Body', *Economy and Society*, 2(1): 70–88.

McCabe, Pete. 2017. *Scripting Magic*. London: Vanishing Inc.

Measom, Tyler and Justin Weinstein. 2014. *An Honest Liar*. Left Turn Films.

Memidex. http://www.memidex.com/contrariwise+to-the-contrary.

Metzner, Paul. 1998. *Crescendo of the Virtuoso: Spectacle, Skill, and Self-Promotion in Paris During the Age of Revolution*. London: University of California Press.

Mitchell, William J. 1994. *The Reconfigured Eye: Visual Truth in the Post-Photographic Era*. Cambridge, MA.: MIT Press.

Mol, Annemarie. 2002. *The Body Multiple: Ontology in Medical Practice*. Durham, NC: Duke University Press. https://doi.org/10.1215/9780822384151.

Morgan, D. 1998. *Focus Groups as Qualitative Research*. London: Sage.

Morris, Errol. 2014. *Believing Is Seeing*. New York: Penguin.

Mulholland, John. 2010. In *The Official CIA Manual of Trickery and Deception*, H. Keith Melton and Robert Wallace (Eds). London: Hardie Grant: 69–81.

Nardi, Peter M. 1988. 'The Social World of Magicians'. *Sex Roles*, 19(11/12): 766.

Neale, Robert E. 2009. 'Early Conjuring Performances', In: E. Burger and R. E. Neale (Eds). *Magic and Meaning* (Second Edition). Seattle: Hermetic Press.

Neale, Robert E. 2008. 'Illusions About Illusions'. In: *Performing Magic on the Western Stage: From the Eighteenth Century to the Present*, Francesca Coppa, Lawrence Hass, and James Peck (Eds). London: Palgrave: 217–230.

Neale, Robert. 1991. *Tricks of the Imagination*. Seattle: Hermetic Press and Jones, G. 2012. 'Magic with a Message', *Cultural Anthropology*, 27(2): 193–214. https://doi.org/10.1111/j.1548-1360.2012.01140.x

Neil, C. L. 1903. *The Modern Conjurer and Drawing-Room Entertainer*. London: C. Arthur Pearson.

Nelms, Henning. [1969] 2000. *Magic and Showmanship*. Mineola, NY: Dover.

Newton, P., Reddy, V. and Bull, R. 2000. 'Children's Everyday Deception and Performance on False-Belief Tasks', *The British Journal of Developmental Psychology*, 2: 297–317. https://doi.org/10.1348/026151000165706.

Noddings, N. 2013. *Caring* (Second Edition). London: University of California Press. https://doi.org/10.1525/9780520957343.

Noyes, P. and Pallenberg, H. 2008. *Women in Boxes: The Documentary Film About Magic's Better Half* [Motion Picture]. http://www.filmbaby.com/films/3277

O'Connor, E. 2005. 'Embodied Knowledge', *Ethnography*, 6: 183–204. https://doi.org/10.1177/1466138105057551.

Olewitz, Chloe. 2020. 'Francis Menotti's Weird Words', *Genii* (November): 39.

Olson, J. A., Landry, M., Appourchaux, K. and Raz, A., 2016. 'Simulated Thought Insertion', *Consciousness and Cognition*, 43: 11–26. https://doi.org/0.1016/j.concog.2016.04.010.

Ortiz, D. 2006. *Designing Miracles: Creating the Illusion of Impossibility*. A-1 MagicalMedia.

Ortiz, Darwin. 1994. *Strong Magic*. Washington, DC: Kaufman & Co.

Owen, Anthony. 2019, April 15. *The Insider*. See https://www.vanishingincmagic.com/blog/the-insider-anthony-owen

Pailhès, A. and Kuhn, G. 2020. 'Influencing Choices with Conversational Primes', *Proc. Natl. Acad. Sci.*, 117: 17675–17679. https://doi.org/10.1073/pnas.2000682117

Pailhès, A. and Kuhn, G. 2020. 'The Apparent Action Causation', *Q. J. Exp. Psychol.*, 73: 1784–1795. https://doi.org/10.1177/1747021820932916.

Palmore, Steve. 2020. *Vanish*, 31: 25.

Palshikar, Shreeyash. 2007. 'Protean Fakir', *Cabinet* (Summer). http://www.cabinetmagazine.org/issues/26/

Paul Draper in 'Scripting Magic 2.1 (Part 2)', 11 September 2020. https://videochatmagic.substack.com/p/scripting-magic-21-part-2

Penn, Jillette and Teller, Raymond. 1992. *The Unpleasant Book of Penn & Teller or How to Play With Your Food*. London: Pavilion.

Penn, Jillette and Teller, Raymond. 1989. *Cruel Tricks for Dear Friends*. New York: Villard Books: 4.

Pettersen, Tove. 2011. 'The Ethics of Care: Normative Structures and Empirical Implications', *Health Care Analysis*, 19(1): 51–64 https://doi.org/10.1007/s10728-010-0163-7

Pickering, Andrew. 2017. 'In Our Place: Performance, Dualism, and Islands of Stability', *Common Knowledge*, 23(3): 381–395. https://doi.org/10.1215/0961754X-3987761.

Pickering, Andrew. 1995. *The Mangle of Practice: Time, Agency, and Science*. Chicago: University of Chicago Press.

Pollner, M. 1987. *Mundane Reason*. Cambridge: Cambridge University Press.

Pritchard, Matt. 2021, September 24. *Comments at SOMA Magic & Creativity Webinar*. https://scienceofmagicassoc.org/blog/2021/8/23/magic-creativity-webinar

Professor Hoffman [Angelo John Lewis], 1876. *Modern Magic*. Eastford, CT: Martino Fine Books. https://www.conjuringarchive.com/list/category/960.

Puig de la Bellacasa, M. 2017. *Matters of Care*. Minneapolis, MN: University of Minnesota Press.

Ragin, Charles and Amoroso, Lisa M. 1994. *Constructing Social Research: The Unity and Diversity of Method*. London: Sage.

Rally, Robert. 2010. *Magic*. Oxford: Oneworld.

Rappert, B. 2021. '"Pick a Card, Any Card": Learning to Deceive and Conceal—With Care', *Secrecy and Society*, 2(2). https://doi.org/10.1177/1468794120965367 and https://brianrappert.net/magic/performances.

Rappert, Brian. 2021. 'Conjuring Imposters'. In: *The Imposter as Social Theory*, Steve Woolgar, Else Vogel, David Moats and Claes-Fredrick Helgesson (Eds). Bristol: Bristol University Press: 147–170.

Rappert, Brian. *The Magic of Social Life*. https://brianrappert.net/the-magic-of-social-life

Regal, David. 2019. *Interpreting Magic*. Blue Bike Productions.

Regal, David. 2021, February 9. *Bristol Society of Magic Lecture*.

Reiter, Sara. 1997. 'The Ethics of Care and New Paradigms for Accounting Practice', *Accounting, Auditing & Accountability Journal*, 10(3): 299–324. https://doi.org/10.1108/09513579710178098

Rissanen, O., Pitkänen, P., Juvonen, A., Kuhn, G., and Hakkarainen, K. 2014. 'Professional Expertise in Magic—Reflecting on Professional Expertise in Magic', *Frontiers in Psychology*. https://doi.org/10.3389/fpsyg.2014.01484

Robert-Houdin, Jean-Eugène. 1859. *Memoirs of Robert-Houdin. Ambassador, Author, and Conjurer*, R. Shelton Mackenzie (Ed.). Philadelphia: George G. Evans.

Roepstorff, A. 2007. 'Navigating the Brainscape'. In: *Skilled Visions: Between Apprenticeship and Standards*, C. Grasseni (Ed.). Oxford: Berghahn Books: 191–206.

Rolfe, Charles. 2014. 'A Conceptual Outline of Contemporary Magic Practice'. *Environment and Planning A: Economy and Space*. 46 (7): 1601–1619.

Rosenthal, Caroline. 2021. 'The Desire to Believe and Belong'. In: *The Imposter as Social Theory* Steve Woolgar, Else Vogel, David Moats, and Claes-Fredrick Helgesson (Eds). Bristol: Bristol University Press: 31–52. ttps://doi.org/10.1332/policypress/9781529213072.003.0001.

Roth, Ben. 2012. 'Confessions, Excuses, and the Storytelling Self: Rereading Rousseau with Paul de Man'. In: *Re-thinking European Politics and History* (Vol. 32), A. Pasieka, D. Petruccelli, B. Roth (Eds). Vienna: IWM Junior Visiting Fellows' Conferences; and

Rouncefield, M. and Tolmie, P. (Eds) 2013. *Ethnomethodology at Work*. London: Routledge. https://doi.org/10.4324/9781315580586.

Ruhleder, K. and Stoltzfus, F. 2000. 'The Etiquette of the Masterclass', *Mind, Culture and Activity*, 7(3): 186–196. https://doi.org/10.1207/s15327884mca0703_06.

Rupar, Aaron. 2019, February 28. 'Trump is "Impressed" that Cohen said "No Collusion." But Cohen Didn't Say that', *Vox*. https://www.vox.com/2019/2/28/18244483/trump-cohen-testimony-vietnam-news-conference-collusion

Saltzman, Benjamin A. 2019. *Bonds of Secrecy*. Philadelphia, Pennsylvania: University of Pennsylvania Press. https://doi.org/10.9783/9780812296846.

Sartre, Jean-Paul. 2003. *Being and Nothingness: An Essay on Phenomenological Ontology*. London: Routledge: 347–348.

Schillmeier, Michael. 2017. 'The Cosmopolitics of Situated Care', *The Sociological Review Monographs*, 65(2): 58. https://doi.org/10.1177/0081176917710426.

Schneider, Tanja and Woolgar, S. 2012. 'Technologies of Ironic Revelation', *Consumption Markets & Culture*, 15(2): 169–189. https://doi.org/10.1080/10253866.2012.654959.

Schutz, A. 1962. *Collected Papers* (Volume 1). The Hague: Martinus Nijhoff.

Schwandt, TA. 1999. 'On Understanding Understanding', *Qualitative Inquiry*, 5(4): 451–464. https://doi.org/10.1177/107780049900500401.

Scott, Susie. 2015. 'Intimate Deception in Everyday Life'. *Studies in Symbolic Interaction*, 39: 251–279. https://doi.org/10.1108/S0163-2396(2012)0000039011.

Shalmiyev, Rich. 2020, June 21. *Presentation in the 'Bridging the Impossible: Science of Magic, Wellbeing and Happiness' Workshop*.

Sharpe, S. H. 2003. *Art and Magic*. Seattle: The Miracle Factory.

Shezam. 2020. *Podcast 54—Catie Osborn on Shakespeare and Tips From an Entertainment Director*. Shezam Podcast. https://shezampod.com/series/shezam/

Shezam. 2019, October 14. *Erik Tait on Publishing*. Magic Podcast 40. https://shezampod.com/podcast/40-erik-tait-on-publishing-magic/

Simons, Daniel J. and Chabris, Christopher F. 1999. 'Gorillas in Our Midst: Sustained Inattentional Blindness for Dynamic Events', *Perception*, 28: 1059–1074.

Singh, Simon. 2003, June 10. 'I'll Bet £1,000 that Derren can't Read my Mind', *The Daily Telegraph*.

Smith, W. et al. (forthcoming). *Explaining the Unexplainable: People's Response to Magical Technologies*.

Smith, W. 2021. 'Deceptive Strategies in the Miniature Illusions of Close-Up Magic'. In: *Illusion in Cultural Practice*, K. Rein (Ed.). Routledge: 123–138.

Smith, Wally. 2016, April 8. 'Revelations and Concealments in Conjuring'. *Presentation to Revelations Workshop*. Vadstena.

Smith, Wally. 2015. 'Technologies of Stage Magic: Simulation and Dissimulation', *Social Studies of Science*, 45(3): 319–343. https://doi.org/10.1177/0306312715577461

Steinmeyer, Jim. 2003. *Hiding the Elephant: How Magicians Invented the Impossible and Learned How to Disappear*. New York: Carroll and Graf.

Suchman, Lucy. 1987. *Plans and Situated Action*. Cambridge: Cambridge University Press.

Sudnow, D. 1978. *Ways of the Hand*. London: MIT Press.

Suzuki, Shunryu. 2005. *Zen Mind, Beginner's Mind*. London: Shambhala.

Tamariz, Juan. 2019. *The Magic Rainbow*. Rancho Cordova, CA: Penguin Magic

Taussig, Michael. 2016. 'Viscerality, Faith, and Skepticism', *Hau: Journal of Ethnographic Theory*, 6(3): 455. https://doi.org/10.14318/hau6.3.033.

Taylor, N. 2018. 'Magic and Broken Knowledge', *Journal of Performance Magic*, 5(1). https://doi.org/10.5920/jpm.2018.03.

Ten Have, Paul. 2004. *Understanding Qualitative Research and Ethnomethodology*. London: Sage. https://doi.org/10.4135/9780857020192.

The Jerk. 2016. 'The Importance of Combining Methods'. http://www.thejerx.com/blog/2016/6/30/the-importance-of-combining-methods

The Jerx. 2015, June 12. 'Presentation Week Part 5: The Distracted Artist Presentation', *The Jerx*.

Thun, Helge. 2019. 'Control', *Genii*. 82(12) December.

Tibbs, G. 2013. 'Lennart Green and the Modern Drama of Sleight of Hand', *Journal of Performance Magic*, 1(1). https://doi.org/10.5920/jpm.2013.1119.

Tolmie, Peter and Mark, Rouncefield. 2013. *Ethnomethodology at Work*. London: Routledge.

Tronto, Joan. 1994. *Moral Boundaries*. London: Routledge.

Tuckett, A. 1988. 'Bending the Truth: Professionals Narratives about Lying and Deception in Nursing Practice', *International Journal of Nursing Studies*, 35(5): 292–302.

Turner, E. 2016. '"I Am Alive in Here": Liveness, Mediation and the Staged Real of David Blaine's Body', *Journal of Performance Magic*, 4(1). https://doi.org/10.5920/jpm.2016.03.

Vernon, Dai. 1940. *Dai Vernon's Select Secrets*. New York: Max Holden

Villalobos, J. Guillermo, Ogundimu, Ololade O. and Davis, Deborah. 2014. 'Magic Tricks'. In: *Encyclopedia of Deception*, Timothy R. Levine (Ed.), Thousand Oaks: Sage: 636–640. https://doi.org/10.4135/9781483306902

Wacquant, Loïc J. D. 2004. *Body & Soul: Notebooks of an Apprentice Boxer*. Oxford: Oxford University Press.

Watzl, Sebastian. 2017. *Structuring Mind: The Nature of Attention and How It Shapes Consciousness*. Oxford: Oxford Scholarship Online. https://doi.org/10.1093/acprof:oso/9780199658428.001.0001.

Weber, Ken. 2003. *Maximum Entertainment*. Ken Weber Productions.

Wenger, E. 1999. *Communities of Practice: Learning, Meaning, and Identity*. Cambridge: Cambridge University Press.

Wetherell, Margaret. 2012. *Affect and Emotion: A New Social Science Understanding*. London: Sage. https://doi.org/10.4135/9781446250945.

Whaley, Barton. 1982. 'Toward a General Theory of Deception' *The Journal of Strategic Studies*, 5(1): 178–192, https://doi.org/10.1080/01402398208437106.

Wieder, D. Lawrence. 1974. 'Telling the Code'. In: *Ethnomethodology*, R. Turner (Ed.). Harmondsworth: Penguin.

Wieder, D. Lawrence. 1974. *Language and Social Reality*. Paris: Mouton.

Youell, Steven. 2009. *Weapons of Mass Deception*. Lecture notes: 45–47.

Zimmerman, Don H. and Pollner, Melvin. 1971. 'The Everyday World as a Phenomenon'. In: *Understanding Everyday Life*, J.D. Douglas (Ed.). London: Routledge & Kegan Paul: 80–103.

Zlatev, Jordan, Brinck, Ingar and Andrén, Mats. 2008. 'Stages in the Development of Perceptual Intersubjectivity', *Enacting Intersubjectivity*. Amsterdam: IOS Press: 117–132.

About the Team

Alessandra Tosi was the managing editor for this book.

Rosalyn Sword performed the copy-editing and proofreading.

Anna Gatti designed the cover. The cover was produced in InDesign using the Fontin font.

Luca Baffa typeset the book in InDesign and produced the paperback and hardback editions. The text font is Tex Gyre Pagella; the heading font is Californian FB. Luca produced the EPUB, AZW3, PDF, HTML, and XML editions — the conversion is performed with open source software such as pandoc (https://pandoc.org/) created by John MacFarlane and other tools freely available on our GitHub page (https://github.com/OpenBookPublishers).

This book need not end here...

Share

All our books — including the one you have just read — are free to access online so that students, researchers and members of the public who can't afford a printed edition will have access to the same ideas. This title will be accessed online by hundreds of readers each month across the globe: why not share the link so that someone you know is one of them?

This book and additional content is available at:

https://doi.org/10.11647/OBP.0295

Donate

Open Book Publishers is an award-winning, scholar-led, not-for-profit press making knowledge freely available one book at a time. We don't charge authors to publish with us: instead, our work is supported by our library members and by donations from people who believe that research shouldn't be locked behind paywalls.

Why not join them in freeing knowledge by supporting us: https://www.openbookpublishers.com/section/104/1

Like Open Book Publishers

Follow @OpenBookPublish

Read more at the Open Book Publishers BLOG

www.ingramcontent.com/pod-product-compliance
Lightning Source LLC
Chambersburg PA
CBHW061250230426
43663CB00022B/2966